Studies in Infrastructure and Control

Series Editors

Dipankar Deb, Department of Electrical Engineering, Institute of Infrastructure Technology Research and Management, Ahmedabad, Gujarat, India

Akshya Swain, Department of Electrical, Computer & Software Engineering, University of Auckland, Auckland, New Zealand

Alexandra Grancharova, Department of Industrial Automation, University of Chemical Technology and Metallurgy, Sofia, Bulgaria

The book series aims to publish top-quality state-of-the-art textbooks, research monographs, edited volumes and selected conference proceedings related to infrastructure, innovation, control, and related fields. Additionally, established and emerging applications related to applied areas like smart cities, internet of things, machine learning, artificial intelligence, etc., are developed and utilized in an effort to demonstrate recent innovations in infrastructure and the possible implications of control theory therein. The study also includes areas like transportation infrastructure, building infrastructure management and seismic vibration control, and also spans a gamut of areas from renewable energy infrastructure like solar parks, wind farms, biomass power plants and related technologies, to the associated policies and related innovations and control methodologies involved.

More information about this series at https://link.springer.com/bookseries/16625

Pravin Jadhav · Rahul Nath Choudhury
Editors

Infrastructure Planning and Management in India

Opportunities and Challenges

Springer

Editors
Pravin Jadhav
Institute of Infrastructure, Technology,
Research And Management (IITRAM)
Ahmedabad, Gujarat, India

Rahul Nath Choudhury
Indian Council of World Affairs
New Delhi, India

ISSN 2730-6453 ISSN 2730-6461 (electronic)
Studies in Infrastructure and Control
ISBN 978-981-16-8836-2 ISBN 978-981-16-8837-9 (eBook)
https://doi.org/10.1007/978-981-16-8837-9

© The Editor(s) (if applicable) and The Author(s), under exclusive license to Springer Nature Singapore Pte Ltd. 2022
This work is subject to copyright. All rights are solely and exclusively licensed by the Publisher, whether the whole or part of the material is concerned, specifically the rights of translation, reprinting, reuse of illustrations, recitation, broadcasting, reproduction on microfilms or in any other physical way, and transmission or information storage and retrieval, electronic adaptation, computer software, or by similar or dissimilar methodology now known or hereafter developed.
The use of general descriptive names, registered names, trademarks, service marks, etc. in this publication does not imply, even in the absence of a specific statement, that such names are exempt from the relevant protective laws and regulations and therefore free for general use.
The publisher, the authors and the editors are safe to assume that the advice and information in this book are believed to be true and accurate at the date of publication. Neither the publisher nor the authors or the editors give a warranty, expressed or implied, with respect to the material contained herein or for any errors or omissions that may have been made. The publisher remains neutral with regard to jurisdictional claims in published maps and institutional affiliations.

This Springer imprint is published by the registered company Springer Nature Singapore Pte Ltd.
The registered company address is: 152 Beach Road, #21-01/04 Gateway East, Singapore 189721, Singapore

Foreword

Infrastructure sector is one of the key drivers of an economy. The sector helps an economy to boost the overall growth and development. It attracts an intense focus from the government and policymakers while formulating any economic policy. Policies would ensure time-bound creation of world-class infrastructure in the country. The infrastructure sector includes power, bridges, dams, roads, and urban infrastructure development. In this modern era, along with the physical infrastructure, digital infrastructure also plays a vital role. In fact, in some sectors digital infrastructure has superseded the requirements of physical infrastructure.

Like any other economy, India realizes the significance of developing a world-class infrastructure. Now, the infrastructure sector has become one of the primary focus areas for the Government of India. India plans to spend US$ 1.4 trillion on infrastructure during 2019–2023 to have a sustainable development of the country. The government has suggested investment of US$ 750 billion for railways infrastructure from 2018 to 2030. Government's deep commitments towards the development of this sector are reflected in its mega projects like Bharatmala Pariyojana and Sagarmala which aim to rejuvenate India's transportation and logistics landscape. The book is a timely publication, considering all the development taking place in Indian infrastructure sectors.

This book attempts to discern the prevailing ideas on the infrastructure policy management system in India. This book, in a very articulate manner, combines the issues faced by different sectors in managing their infrastructure facilities. This book is an attempt to identify some of the critical risks and challenges in the planning and management of infrastructure in India. This book analyses the diverse management solutions that can offer support for better infrastructure management across sectors.

I sincerely believe that this book will be an invaluable addition to the existing literature on infrastructure management in India. It promises to be of major benefit to readers with a general interest in this topic. Policymakers will also profit from in-depth research on infrastructure management and related policies. I am confident that this book will greatly serve as a ready reference for the students, academics, researchers, and policymakers. Plaudits are owed to the editors, Dr. Pravin Jadhav and Dr. Rahul Nath Choudhury, for their efforts in producing this comprehensive volume and offering such an in-depth analysis of a topic of such immense importance. It is my hope that the authors will continue to work on the many and varied aspects of infrastructure management in India as well as in other countries. I wish them every success in their future efforts, including contributions to the study and examination of these and related issues to the benefit of all concerned.

Rajan Sudesh Ratna
Deputy Head and Senior Economic
Affairs Officer, United Nations ESCAP
South and South-West Asia Office
New Delhi, India

Acknowledgements

The editing process for this book began in October 2020. Because this is our first edited book, we were excited when we received the publisher's final clearance. Without our close association with a number of people, this edited book would not have been possible. We would like to take this opportunity to express our heartfelt gratitude and appreciation to everyone who helped make our first edited book a reality.

First and foremost, we would like to express our sincere gratitude to the Institute of Infrastructure Technology Research and Management (IITRAM) for giving us all the necessary support and resources to work on this edited book. We are very much thankful to Prof. Shiva Prasad, Director General, IITRAM, for his inspiring mentorship. We are also thankful to Dr. A. U. Digraskar, Director, IITRAM, for continuous support and advices which have greatly helped towards the successful organization of this event.

We would like to express our deep gratitude to Dr. Deepankar Deb, Series Editor, Studies in Infrastructure and Control, for overall technical guidance from time to time to improve the quality of book. We are grateful to the anonymous reviewers for keeping their faith on us since the proposal stage and pushing us to complete the task.

We would like to express our heartfelt gratitude to all of the authors for their contributions to this edited volume. We appreciate their timely submission of their chapters.

Last but not least, we extend our sincere thanks to almighty God and others who have helped us directly or indirectly at every stage to complete this edited book.

Introduction

Importance of Quality Infrastructure

Infrastructure plays a crucial role in any nation's economic development as it increases the nation's productivity and competitiveness by decreasing the overall cost of production. Infrastructure growth also leads to inclusive growth of a nation. One of the primary aspirations of any country and its citizen is to have a good quality of roads for better transportation, a home where electricity under no circumstances fails, adequate tap water supply for each family, good quality of schools, hospitals, parks, etc. Availability of superior quality of physical infrastructure can also increase the rate of return from the investment. Ushering the changing requirements of individuals and businesses, nations have shifted their priority towards developing infrastructure. Along with the physical infrastructure, economies have started investing heavily on the digital facilities. In a nutshell, infrastructure plays a significant role in choosing the destination by multi-national corporations (MNCs) while making overseas investment decisions. Modern infrastructure combined with efficient management is a key to business and economic development. Sahoo and Dash argue that infrastructure development plays a crucial role in the economic growth of India [6]. A large body of the literature claims that infrastructure development increases economic activities and production facilities while reducing the trade and production cost. According to Pradhan and Bagchi, rail and road transportation infrastructure enhanced India's economic growth [5]. Most of these studies suggested that policy intervention is needed to boost infrastructure development and sustainable economic growth in India. Global Competitiveness Report, 2019, publishes annually by the World Economic Forum arguing that infrastructure is the second important pillar for improving its competitiveness [8].

Infrastructure is a crucial component while evaluating a nation's competitive advantage by the foreign investor. Like India, most of the world's developing countries are consistently dynamic and proactive to create a favourable policy regime to attract more infrastructure investment. Being one of the fastest growing developing countries with several states, India has accorded continuously greater significance to the infrastructure division and gained extensive consideration from the legislature and private venture capitalists. An enormous spotlight has consistently accorded on accomplishing allied ventures employing public–private partnerships (PPPs), financial impetuses, tariff schemes, and fiscal incentives. Indian infrastructure segment mostly incorporates the improvement of roads, air terminals, transporting, and ports that have significantly supported the Indian economy in the course of the most recent years. However, India is not able to create significant development in the infrastructure sector. India's government plans to spend the US $ 1.4 trillion on infrastructure during 2019–2023 as a part of the National Infrastructure Pipeline (NIP). The NIP is India's attempt to bring private sector for forging partnerships via effective models for co-working between public and private sectors. India recognizes that lack of robust infrastructure is a serious impediment to the development of an economy[1]. In addition, the outcome of India's ambitious Make in India plan to revive the manufacturing sector depends largely on the development of required infrastructure. There is a constant need for government intervention, solid funding, and constant monitoring of projects. Better planning and management are crucial to sustainable infrastructure development. Realizing the importance of the infrastructural development, India allocated a budget of Rs. 233,083 crore (US$ 32.02 billion) to enhance the transport infrastructure 2021, in the Union Budget 2020–2021. Further, the government expanded the NIP to 8158 projects from 7400 projects.

Necessity of Infrastructure Management

Having developed the standard infrastructure facilities, the challenge is to maintain the same with proper management and efficient planning. In transport infrastructure, well-managed and functional facilities such as airports and seaports enable us to connect with the other parts of the world; an efficient IT and digital infrastructure equips us with advance communication system, while schools and hospitals ensure well-being, including a healthy and well-equipped workforce. Better infrastructure facilities are also necessary to have higher productivity. Studies have established the positive link between higher infrastructural spending and economic development in a country. Scholars have estimated that infrastructure resources must be well spent to achieve high-quality infrastructure. The World Bank, 1994, finds that higher infrastructural spending leads to the higher GDP growth. However, the growth varies with countries [7]. Aschaue, studying samples from the countries like UK, Germany, Japan, Mexico, Sweden, and the USA, concludes that higher infrastructural spending positively impacts the productivity growth in a nation [1]. Some other

---

[1] National Infrastructure Pipeline. Report of the Task force, Department of Economic Affairs. Ministry of Finance. Government of India. Available at: https://dea.gov.in/sites/default/files/Report%20of%20the%20Task%20Force%20National%20Infrastructure%20Pipeline%20%28NIP%29%20-%20volume-i_1.pdf Accessed on 17.9.21.

important studies are making similar arguments [2,3,4]. We need to develop mechanisms that ensure that infrastructure is of high quality and is sustainable over the long run. Further, with the rapid increase in urbanization and population, there is greater demand for adequate and sustainable infrastructure in transportation, electricity, water and sanitation, social and economic infrastructure, etc. Therefore, planning and management of infrastructure are the vital policy agendas for any nation.

About This Book

Realizing the importance of the subject, we are making an attempt to combine ideas from some of the established scholars and practitioners in various fields of infrastructure policy and management. This book presents those ideas into a structured and systematic manner. This book is an attempt to contribute to this sector by addressing the opportunities and the challenges faced by the industry. The primary objective of the book is to critically analyse sectoral Infrastructure planning and management from an Indian perspective. As discussed above, it is imperative to understand different planning and management practices, their significance, and issues to have sustainable infrastructure development. This book tries to identify empirical risks and challenges in the planning and management of infrastructure in India. This book also analyses diverse management solutions that can support better infrastructure management across sectors and critically discuss the different ways to overcome these risks related to infrastructure planning and management in India. This book also covers real-world case studies related to infrastructure management from various stakeholders involved in the management of infrastructure.

This book covers four crucial sectors, viz. transportation, power, smart cities, and IT/digital infrastructure. Accordingly, this book has four sections. There are eleven chapters in the book of which four are under the first section covering transport infrastructure. In the second section of the book, we have three chapters discussing various issues in the energy and power infrastructure sector. Third section of the book explores the three important urban infrastructure issues, while the final section focuses on the growing digital infrastructure in India.

In the existing literature, there is no book available that has attempted to address comprehensive issues of infrastructure management at the sectoral level in India. The available books have discussed infrastructure management at the macro-level or focusing on only one sector. Hence, this is an essential book considering the lacuna in the literature. Sustainable infrastructure is very much crucial for world development. Therefore, the current research is critical in this perspective. The book will be a primary and pioneering attempt to address this vital issue.

Prelude to the Chapters

Chapter 1 of the book discusses the road infrastructure system in India focusing on the road safely and road transport management in India. The chapter is contributed by Jiten Shah, Khushbu Bhatt, Priyank Trivedi, and Said Easa. The authors observe that due to poor road safely measures, there has been a manifold increase in road accidents in India. The authors find India among the worst-performing countries in the world in terms of road safety measures. The chapter identifies and analyses the challenges of potential hazards related to the road transport safely. The authors make an assessment of the existing safety measures and their performance. The chapter suggests a set of remedial measures which would be helpful to make the road transport system safer. The authors argue that these proposed measures will be helpful for policymakers in the implementation of road safety, infrastructure, and its management in India.

Chapter 2 titled 'Port Development: History, Present and Future Challenges' is articulated by Abhijit Singh. As the title suggests, this chapter offers a scholarly narration of the history about Indian port along with depicting the current scenarios of the same. The author also points out the crucial challenges that exist in the Indian port and its management practice. Analysing the port management in India from various perspectives, the author comments that ports are far from satisfactory level in terms of their efficiency, and the ports consume much higher span of time in custom clearance and many other trade-related issues compared to other ports in the region. Infrastructures of Indian ports are also in a very poor stage. The chapter stresses the need to upgrade the port system with digital infrastructure and modern technology to bring efficiency in the port management.

Taking the discussion forward about the transport infrastructure, the third chapter written by Manish Yadav and Tarun Dhingra enlists the importance of airport infrastructure in the Indian economy enrooted from operational, planning, and management outlook. This chapter describes the operational aspect of airport, including classification, i.e. kerbside, terminal, and airside operations. The authors attempt to pen down the fundamentals of airport infrastructure planning, including capacity planning, master planning, facility and layout planning. They also illustrate the management perspective of airport infrastructure and evaluate commercial revenues, including aeronautical and non-aeronautical revenues with financial sustainability. The chapter also explores the public–private partnerships and its impact on airport competition, growths, and sustainability contributing to the Indian economy and its sustainability in the aviation sector.

The final chapter of the section authored by Tarun Dhingra and Sanjeev Sharma outlines the journey of Indian railway. Indian railway is the largest cargo and passenger service providers in India at the lowest price. It is also the biggest public sector employer in the country. Indian railway has got immense economic importance because of these features. In this chapter, the authors attempt to review the current status and the important milestones achieved by Indian Railway since the independence. The chapter finds maintaining the lower rate of tariffs is the major challenge for Indian Railway at the current stage. Due to other modes of transport at a reasonable price, railways are also struggling to maintain a high number of passenger. The authors suggest, to remain competitive vis-à-vis other transportation modes and provide optimum service to passengers/freight, Indian Railways needs to upgrade its current infrastructure. Greater private sector engagement in this sector by creating a conducive policy and regulatory environment is the need of the hour.

Part II of the book inquires about the management of Indian power and energy infrastructure. This section starts with the debate on issues and challenges in the power sector infrastructure management in India. The first chapter in this section is contributed by Anil Kumar and Avishek Ghosal. In this chapter, the authors argue that Indian power sector is in transition phase and shifting the focus from conventional resources to non-conventional resources. Several measures have taken to bring reforms in this sector, and revival of the power distribution infrastructure is underway. This chapter tactfully analyses various issues of concern in this sector along with the various policy variables. Policies undertaken in enhancing competition in power market are some of the highlights of this chapter. The chapter also extends suggestions to bring reforms to ensure a secured, affordable, and reliable sustainable energy future of India.

Nikita Das in Chap. 6 explores the status quo of Indian renewable energy management system. In this chapter, the authors highlight the governmental, institutional, and regulatory infrastructures in place for renewable energy management in India. The chapter makes further attempts to identify the various issues and challenges that pertain to such management and proposes suggestions to circumvent the same. The author argues that renewable energy cannot be deemed as a silver bullet at the outset, but we can pave the way for establishment of meaningful infrastructure and its effective management in the long run. Research and development to introduce new green technologies as alternatives to fossil fuel-based technologies must be considered in tandem with pushing for economies of scale that can make these technologies competitive.

The section of the book ends with a chapter articulated by Shailly Kedia and Nivedita Cholayil. This chapter assesses the issue of climate change and its linkages with the sustainable development in India. This chapter critically examines recent narratives in India on energy in the context of climate change and sustainability. They argue that attaining energy security is essential for human development and economic growth. The chapter finds the discourse on sustainability at the domestic level in India is also very much driven by diversifying fuel share in electricity mix and not diversification in primary energy mix. They also comment that energy security still remains crucial as India relied on imports for about 40% of fuel needs in terms of primary energy.

India has witnessed massive growth in its urban population over the last few decades. Governments and policymakers are facing challenges such as increasing urban population and huge infrastructure gaps to meet various demands. To provide modern and better facilities and to raise the quality of life in Indian cities, the concept of smart city has been devised. The concept of smart city is increasingly considered a new paradigm of sustainable growth. To understand this issue in depth, Chap. 8 of this book focuses on the key issues and the challenges faced in developing new cities or improves the infrastructure facilities in existing cities in India. This chapter is contributed by Vinay Kandpal, Vikas Tyagi, and Harmeet Kaur.

With the development of the city infrastructure, the modes of transportation are also changing. A transition is noticed in the vehicles. Electric vehicles are replacing the old petrol- and diesel-run vehicles. They are transforming the mobility experience across the world. Electric vehicle is considered to be sustainable mode of transportation. It consumes less energy and reduces less carbon. India has potential to achieve electric mobility future by utilizing existing conditions, and government programmes and policies. To upscale adoption of electric vehicles and for its management in India, issues such as charging infrastructure, research and development, financing of electric vehicles, battery and cell manufacturing, and proper regulatory framework need to be revamped immediately. Shikha Juyal analyses this vital issues in Chap. 9 of Section III. She tries to explain that the electric mobility pathway would provide clean, low-cost mobility, create new jobs, reduce oil imports, improve health of people, and would have positive economic impact. The chapter highlights the policies and number of incentives provided by the Government of India. Finally, she outlines the major challenges which need to be addressed to boost adoption of electric vehicles in future in India.

The last chapter in the section authored by Muhammadriyaj Faniband, Kedar Vijay Marulkar, and Pravin Jadhav explores the issues of the sustainable development Indian context. They examine the drivers and the barriers to sustainable development in India. This chapter reviews multiple projects launched in India to bring sustainable development in various sectors.

The final section of the book focuses on the digital infrastructure facilities and their management. India has been in the forefront in developing digital infrastructure system. Several policies have been devised to develop and popularize digital services ranging from the financial sector to health sector. Public schemes such as Digital India and creating National Optical Fibre Network are some of the commendable steps taken in India. Stressing the need of developing a world-class digital infrastructure, Krishna Teja Perannagari and Vineet Gupta in Chap. 11 discuss India's digital transformation, highlighting various programmes and initiatives taken by the Indian government to foster the goal of Digital India. They also shed light on recent technological trends influencing the development of digital infrastructure and suggest measures to encourage the adoption of latest technology and promote investments in digital infrastructure projects.

The book ends with an editorial summary and conclusion.

Rahul Nath Choudhury
Pravin Jadhav

References

1. Aschauer, D. A. (1989). Is public expenditure productive? *Journal of Monetary Economics*, *23*, 177–200.
2. Berndt, E. and B. Hanson. (1992). Measuring the contribution of public infrastructure capital in Sweden. *The Scandinavian Journal of Economics*, *94*, 151–S168.
3. Conrad, K. and H. Seitz. (1994). The economic benefits of public infrastructure. *Applied Economics*, *26*, 303–311.
4. Nadiri, I. and T. Mamuneas. (1994). The effects of Public Infrastructure and R&D capital on the cost structure and performance of U.S. manufacturing industries. *Review of Economics and Statistics*, *27*, pp. 22–37.
5. Pradhan, R. P., & Bagchi, T. P. (2013). Effect of transportation infrastructure on economic growth in India: The VECM approach. *Research in Transportation Economics*, *38*(1), 139–148. https://doi.org/10.1016/j.retrec.2012.05.008.
6. Sahoo, P., & Dash, R. K. (2009). Infrastructure development and economic growth in India. *Journal of the Asia Pacific Economy*, *14*(4), 351–365. https://doi.org/10.1080/13547860903169340
7. World Bank (1994). The World Development Report: Infrastructure and Development. The World Bank. Washington D.C.
8. World Economic Forum (2019) Global Competitiveness Report, 2019. World Economic Forum.

Contents

Part I Transport Infrastructure: Fostering the Economic Growth

1 Road Safety Conditions and Management in India: Challenges and Opportunities 3
Jiten Shah, Khushbu Bhatt, Priyank Trivedi, and Said Easa

2 Port Development: History, Present and Future Challenges 25
Abhijit Singh

3 Emerging Scope of Airport Infrastructure: Case of India 35
Manish Yadav and Tarun Dhingra

4 Rail Infrastructure—Journey Since Indian Independence and Beyond 53
Tarun Dhingra and Sanjeev Sharma

Part II How Power Sector is Managed in India?

5 Power Sector Infrastructure Management: Issues and Challenges 91
Anil Kumar and Avishek Ghosal

6 Renewable Energy Management: An Analysis of the Status Quo 99
Nikita Das

7 Energy, Climate Change and Sustainable Development in India ... 129
Shailly Kedia and Nivedita Cholayil

Part III The Emergence of Modern Cities: Smart or Sustainable?

8 Smart City: Sustainable City for Tackling Urban Challenges 147
Vinay Kandpal, Vikas Tyagi, and Harmeet Kaur

9	**Electric Mobility and Electric Vehicles Management in India** ...... Shikha Juyal	159
10	**Sustainable Infrastructure Development in India: Drivers and Barriers** .... Muhammadriyaj Faniband, Kedar Vijay Marulkar, and Pravin Jadhav	173

Part IV Developing Digital Infrastructure

11	**Recent Trends in Digital Infrastructure in India** .... Krishna Teja Perannagari and Vineet Gupta	187

Conclusion 203

Editors and Contributors

About the Editors

Dr. Pravin Jadhav obtained Ph.D. from the Indian Institute of Foreign Trade (IIFT) under the Ministry of Commerce and Industry, Government of India in 2015. Dr. Jadhav was associated with IIFT from January 2008 to July 2012. In IIFT, he did extensive research and undertook various research studies for the Ministry of Commerce and Industry, Ministry of Science and Technology, and the European Union. He was also served as a researcher in the Planning Commission working Sub-Group on Technology Intensity in India's Manufacturing Exports, Planning Commission for framing the 12th Five Year Plan (2012–2017).

Presently, he has been working as an assistant professor at the Institute of Infrastructure, Technology, Research And Management (IITRAM), an Autonomous University established by the Government of Gujarat. Currently, he is working in the area of infrastructure planning and management.

Dr. Rahul Nath Choudhury is a trade economist currently associated with the Indian Council of World Affairs, New Delhi as a Research Fellow. His primary research interest includes foreign direct investments, international trade, political economy, geo-economics, and digital trade. Rahul has diverse experience of working in both the public and the private sector in academia and the industry in various capacities. He has contributed to numerous book chapters and academics journals. Rahul has been a freelance consultant for organizations like ADB, IFC, AEPC, CII, and a few MNCs as well.

Contributors

Khushbu Bhatt Department of Civil Engineering, Institute of Infrastructure, Technology Research and Management, Ahmedabad, Gujarat, India

Nivedita Cholayil The Energy and Resources Institue, Darbari Seth Block, India Habitat Centre, New Delhi, India

Nikita Das Department of Anthropology, University at Buffalo, SUNY, Buffalo, USA

Tarun Dhingra School of Business, University of Petroleum & Energy Studies, Dehradun, India

Said Easa Department of Civil Engineering, Ryerson University, Toronto, ON, Canada

Muhammadriyaj Faniband Christ Academy Institute for Advanced Studies, Bengaluru, India

Avishek Ghosal University of Petroleum and Energy Studies, Dehradun, India

Vineet Gupta School of Business Studies, Sharda University, Noida, India

Pravin Jadhav Institute of Infrastructure, Technology, Research And Management (IITRAM), Ahmedabad, India

Shikha Juyal NITI Aayog, New Delhi, India

Vinay Kandpal School of Business, Department of General Management, UPES, Dehradun, India

Harmeet Kaur Management Department, Jharkhand Rai University, Ranchi, India

Shailly Kedia The Energy and Resources Institue, Darbari Seth Block, India Habitat Centre, New Delhi, India

Anil Kumar University of Petroleum and Energy Studies, Dehradun, India

Kedar Vijay Marulkar Department of Commerce and Management, Shivaji University, Kolhapur, India

Krishna Teja Perannagari MICA - The School of Ideas, Ahmedabad, India

Jiten Shah Department of Civil Engineering, Institute of Infrastructure, Technology Research and Management, Ahmedabad, Gujarat, India

Sanjeev Sharma FA&CAO, Northern Railway, New Delhi, India

Abhijit Singh Executive Director, IPA (An Apex Body of Major Ports Under Administrative Control of Ministry of Ports, Shipping and Waterways, Govt. of India), New Delhi, India

Priyank Trivedi Department of Civil Engineering, Institute of Infrastructure, Technology Research and Management, Ahmedabad, Gujarat, India

Vikas Tyagi Department of Management, Chandigarh University, Ajitgarh, India

Manish Yadav Department of Advanced Aviation Management, Qatar Aeronautical College, Doha, Qatar

Abbreviations

5G	Fifth Generation
AASHTO	American Association of State Highway and Transportation Officials
ACC	Advanced Cell Chemistry
AEPS	Aadhaar Enabled Payment System
AERA	Airport Economic Regulatory Authority
AI	Artificial Intelligence
AIIB	Asian Infrastructure Investment Bank
AMP	Automotive Mission Plan
ANS	Air Navigation Services
AOCC	Airport Operation Control Center
APB	All-points Bulletin
API	Application Programming Interface
ARFF	Aircraft Rescue and Fire Fighting
BEE	Bureau of Energy Efficiency
BOT	Built, Operate and Transfer
CEA	Central Electricity Authority
CERCs	Central Electricity Regulatory Commissions
CISF	Central Industrial Specific Force
CNS	Communication, Navigation and Surveillance
CSIA	Chhatrapati Shivaji International Airport
DDUGJY	Deen Dayal Upadhyaya Grameen Jyoti Yojana
DGCA	Directorate General of Civil Aviation
DHI	Department of Heavy Industry
DIKSHA	Digital Infrastructure for Knowledge Sharing
DISCOMs	Electricity Distribution Companies
DSM	Deviation Settlement Mechanism
DVB	Delhi Vidyut Board
EBP	Ethanol Blending in Petrol
EC	European Commission
ECBC	Energy Conservation Building Code
eKYC	Electronic Know Your Customer

EoI	Expression of Interest
eSign	Electronic Signature
EV	Electrical Vehicle
EVSE	Electric Vehicle Supply Equipment
EXIM	Export-Import Bank of India
FAME	Faster Adoption and Manufacturing of Hybrid and Electric Vehicles
FDI	Foreign Direct Investment
FOD	Foreign Object Debris
FOWIND	Facilitating Offshore Wind in India
GDP	Gross Domestic Product
GHG	Greenhouse Gas
GNI	Gross National Income
GOI	Government of India
GTAM	Green Term Ahead Market
GW	Gigawatt
HSM	Highway Safety Manual
HUDCO	Housing and Urban Development Corporation
IATA	International Air Transport Association
ICE	Internal Combustion Engine
IDBI	Industrial Development Bank of India
IDF	Infrastructure Debt Fund
IEA	International Energy Agency
IFC	International Finance Corporation
IFCI	Industrial Finance Corporation of India
IFCs	Infrastructure Finance Companies
IGA	Intergovernmental Agreement
IGIA	Indira Gandhi International Airport
IHME	Institute of Health Metrics and Evaluation
IIFCL	India Infrastructure Finance Company Limited
IIPDF	India Infrastructure Project Development Fund
ILS	Instrumental Landing System
IMPS	Immediate Payment Service
INDC	Intended Nationally Determined Contributions
IoT	Internet of Things
IPA	Indian Port Association
IPCC	Intergovernmental Panel on Climate Change
IPDS	Integrated Power Development Scheme
IR	Indian Railways
IRAP	International Road Assessment Program
IRC	Indian Road Congress
IREDA	Indian Renewable Energy Development Agency Ltd
ISRO	Indian Space Research Organisation
IT	Information Technology
LCC	Low-Cost Carrier
LED	Light-Emitting Diode

LNG	Liquefied Natural Gas
MB	Mega Byte
MCAs	Model Concession Agreements
MCT	Minimum Connecting Time
MDB	Multilateral Development Bank
MERC	Maharashtra Electricity Regulatory Commission
MNRE	Ministry of New and Renewable Energy
MOAE	Ministry of Atomic Energy
MOC	Ministry of Coal
MoCA	Ministry of Civil Aviation
MoEFCC	Ministry of Environment, Forest and Climate Change
MOHIPE	Ministry of Heavy Industries and Public Enterprises
MOP	Ministry of Power
MOPNG	Ministry of Petroleum and Natural Gas
MORTH	Ministry of Road Transport and Highways
MOSPI	Ministry of Statistics and Programme Implementation
MTOE	Million Tonnes of Oil Equivalent
MYTP	Multi-Year Tariff Planning
NABARD	National Bank for Agriculture and Rural Development
NAPCC	National Action Plan on Climate Change
NBFC	Non-Bank Finance Company
NCEF	National Clean Energy Fund
NDB	National Development Bank
NEC	National Economic Council
NEMMP	National Electric Mobility Mission Plan
NHAI	National Highway Authority of India
NIIF	National Investment and Infrastructure Fund
NIP	National Infrastructure Pipeline
NIWE	National Institute of Wind Energy
NMT	Non-Motorized Transport
NSGM	National Smart Grid Mission
NSM	National Solar Mission
NTKMs	Net Tonne Kilometres
OA	Open Access
OECD	Organisation for Economic Co-operation and Development
OEMs	Original Equipment Manufactures
OTP	On-Time Performance
PCIC	Per Capita Investment Cost
PCS	Port Community System
PFC	Power Finance Corporation Ltd
PGCIL	Power Grid Corporation of India Limited
PIARC	Permanent International Association of Road Congress
PKMs	Passenger Kilometres
PMP	Phased Manufacturing Program
PPP	Public–Private Partnership

PV	Solar Photovoltaic
RAP	Road Assessment Program
RBI	Reserve Bank of India
RE	Renewable Energy
REC	Rural Electrification Corporation
RECs	Renewable Energy Certificates
ROT	Runway Occupancy Time
RPO	Renewable Purchase Obligation
RRECL	Rajasthan Renewable Energy Corporation Limited
RSI	Road Safety Inspection
RSM	Road Safety Management
RTM	Real-Time Market
RWSS	Rural Water Supply and Sanitation
SC	Smart City
SDG	Sustainable Development Goal
SEB	State Electricity Board
SEBI	Securities and Exchange Board of India
SECI	Solar Energy Corporation of India
SERCs	State Electricity Regulatory Commissions
SIDBI	State Industrial Development Banks
SLNP	Street Lighting National Programme
SMEV	Manufacturers of Electric Vehicles
SPaRC	Solar Power as a Remunerative Crop
TERI	The Energy and Resources Institute
TUFC	Terminal User Facility Charges
UDAY	Ujwal DISCOM Assurance Yojana
UJALA	Unnat Jyoti by Affordable LEDs for All
UNO	United Nations Organization
UPI	Uniform Payment Interface
VGF	Viability Gap Funding
VNR	Voluntary National Review
WHO	World Health Organization

Part I
Transport Infrastructure: Fostering the Economic Growth

Chapter 1
Road Safety Conditions and Management in India: Challenges and Opportunities

Jiten Shah, Khushbu Bhatt, Priyank Trivedi, and Said Easa

Abstract Due to the lack of proper roads, transport infrastructure, and adequate governance, there has been a manifold increase in road collisions in India. India is among the worst-performing countries in terms of road safety and witnesses the highest number of collisions in the world every year. This chapter identifies and analyzes two primary groups of challenges of road safety in India: (1) challenges of road and traffic conditions and (2) challenges of the road safety management (RSM) system. For road and traffic conditions, the traffic collision problems due to road and traffic characteristics from the engineering perspective are discussed, addressing the effects of speed, geometric design, and driving behavior on road safety. The RSM challenges are grouped into three categories: (a) management framework and policy challenges, (b) engineering design and standardization challenges, and (c) awareness challenges. Based on these challenges, several strategies related to infrastructure management, standards provision, policymaking, and enforcement of laws and regulations are suggested to help policymakers in implementing road safety management in India.

Keywords Road · Collision · Transport · Safety · Transportation management

1.1 Introduction

Collisions kill more people in India than terrorism and natural disaster [8]. During the past decade, the natural cause of collisional death was lightning accounts for 10%. However, among the unnatural causes, traffic collisions were recorded as 42–45% of the unnatural deaths. As per the report of Institute of Health Metrics and Evaluation, in 2016, road traffic injuries were the 8th leading cause of death in India

---

J. Shah (✉) · K. Bhatt · P. Trivedi
Department of Civil Engineering, Institute of Infrastructure, Technology Research and Management, Ahmedabad, Gujarat, India
e-mail: jitenshah@iitram.ac.in

S. Easa
Department of Civil Engineering, Ryerson University, Toronto, ON, Canada

Table 1.1 Road collision statistics in India from 2016 to 2018 [19]

Parameter	2016	2017	2018
Number of road collisions	4,80,652	4,64,910	4,67,044
Number of persons killed	1,50,785	1,47,913	1,51,417
Number of persons injured	4,94,624	4,70,975	4,69,498
Collision severity (fatalities/100 collisions)	31.4	31.8	32.4

in which the health loss among the young (15–49 years) is quite significant with the proportion of 68.7%, according to the Ministry of Road Transport and Highways [19]. According to the World Health Organization, road traffic injuries caused an estimated 1.35 million deaths worldwide in the year 2016. However, India ranks first among the 199 countries for road collisions as per world road statistics 2018. The data for road collisions for the last three years have been enumerated in Table 1.1.

Among the different categories of roads, more severe collisions are reported on rural roads. Kadiyali et al. [16] found that the leading cause of collisions on rural roads is the geometry, road width, traffic volume, and different intersections. Singh [37] and Mirkazemi [20] suggested local safety improvements suggested to reduce the traffic collisions at highways, such as correction of geometric curves, provision of paved shoulders, placing guard rails, removal of roadside hazards traffic barriers, and advanced warning systems. As per the statistics [19], in India, collisions on low-class roads have not been registered as frequent hit and run cases. This information suggests improving the road features, educating road users, and improving collision reporting to obtain the actual fatal and injuries to improvise the conflicts. One of the case studies by Pandey et al. [26] of Vishakhapatnam specified the attributes for the collision were lack of signage, obstruction due to trees and bushes, improper visibility for pedestrians and drivers, median openings, and uncontrolled access for wrong side movements at NH 5. Moreover, the collision rate for a different classification of vehicles was noted as 35% for two-wheelers, 23% for goods vehicles, followed by cars, autos, and buses as 17%, 15%, and 9%, respectively. The authors suggested National Highway Authority of India (NHAI) close the median openings, remove the encroachments and improve the intersection geometrics at National Highways.

Sarin et al. [34] suggested cost-effective improvements to reduce collisions and increase safety in the USA and UK. The improvements included road and roadside and operational improvements by providing traffic signs, markings, minor alterations in the layout of intersections, and drainage improvements. Other finding shows that front to rear collision involves buses and heavy vehicles due to slowing down or the stopping or overtaking which accounts for 59% of the collisions, which can be mitigated by implementing acceleration and deceleration lanes, warning and information signs. Bashar et al. [4] examined 76 collisions on National Highways, where 58% were at intersections. Besides the shift of collision pattern from a head-on collision on the two-lane undivided highway to the front-rear collision on a four-lane divided highway. Proper signage facilities and advance warning systems need to be introduced, and the appropriate design and location of intersections and median openings to diminish the collisions. Based on random parameter modeling for two-lane undivided highways in India, the factors influencing the collisions at the daytime and nighttime are compared, which illustrates the increase in the proportion of cars and shoulder width decreases the collision frequencies at day time. In contrast, the proportion of cars and trucks increases, the rate of collision decreases in the nighttime by Raichur et al. [30].

A road safety audit conducted by Jha et al. [15] suggested providing a service road on highways to bifurcate the fast- and slow-moving traffic that creates hazards. They also suggested developing adequate lighting for all major and minor intersections at which many collisions were observed. A road safety report [19] on two-lane single carriageway rural networks in four states (Karnataka, Andhra Pradesh, Gujarat, and Assam) recommended upgrading roundabouts, signalized intersections, and turning lanes, and upgrading road features, like paving and widening shoulders. The combined investment for all four states will prevent 1.25 million deaths and serious injuries, reducing 40% deaths based on the estimates implemented by Kanchan et al. [17].

The study by Singh and Mishra [38] at NH 6, the collision rate can be reduced by 50% by increasing the shoulder width by 1 m as the collision rate of 1.62 collisions per year per km and additional access point per km per road length may increase the road collisions, especially for the pedestrians. An investigation executed by Dinu and Veeraragavan [7] on 167 collisions was carried out in Coimbatore (Tamil Nadu), included 71 fatal collisions involving 80 fatalities (66 vehicle occupants and 14 pedestrians) which accounts for 43% of all collisions investigated and the injury severity MAIS from 3–6 and fatalities MAIS is 4. Median opening and access to roads significantly influence collisions as per the realistic experience carried by Prajapati and Tiwari [29]. The buses are involved in about 10–35% of the road collisions in different states of India, where the collision rates for straight roads, curves, and intersections for buses are 81%, 11%, and 5%, respectively. Singh [36] and Kar et al. [18] recommended road safety education and practical driver training to reduce the collision rate.

In developing countries like India, the rate of collisions is significant compared to other countries (Fig. 1.1). Moreover, motor vehicles have increased from 1401 to 230,031 (in thousands), and the collisions increased from 114 to 464 (in thousands) during 1970–2017. The national average collision severity (persons killed per 100 collisions) increased from 28.5 in 2014 to 31.5 in 2017. Various preventive measures have been recommended, including lane bifurcation, pedestrian safety, traffic rules regulations, enforcement, and education suggested by Ratanavaraha and Suangka [32].

Moreover, globally, the death rate due to collisions is increasing. The primary cause of the collision is speed, not using preventive measures like helmets or seat belts, and drinking and driving. Therefore, safety measures are taken into account by improving the road infrastructure and design standards as per the laws in 112 countries worldwide to reduce road collisions. The fatality rates for different countries, along with the growth of vehicles and fatality risk, are shown in Table 1.2. As noted, the fatality rate and risk are higher in India (8.6 and 11.2, respectively) are the highest,

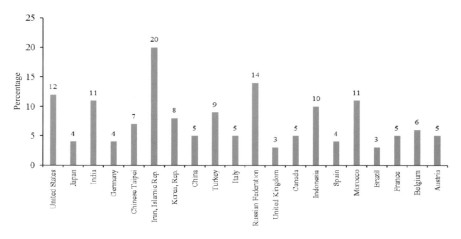

Fig. 1.1 Country-wise number of fatalities per million population in 2018 [19]

Table 1.2 Comparison of fatalities in 2012 for different countries [37]

Country	Motorization rate (no. vehicles/10^3 people)	Fatality rate (no. fatalities/10^4 vehicles)	Fatality risk (no. fatalities/10^5 people)
India	130	8.6	11.2
Germany	657	0.67	4.4
Japan	651	0.63	4.1
New Zealand	733	0.91	6.9
Sweden	599	0.50	3.0
UK	599	0.51	2.8
USA	846	1.26	10.7

and the motorization rate (130) is the lowest. In contrast, the fatality rate and risk in the USA are (1.26 and 10.7, respectively), and the motorization rate (846) is the highest. This is followed by New Zealand that has a fatality rate, fatality risk, and motorization rate of 0.91, 6.9, and 733, respectively. Hence, the roads in Indian are less safe than those in other developed countries.

Various collision studies have been conducted in developed and developing countries considering various statistical analyses and different variables. The studies predict the severity of the collision, collision trends, factors affecting the collision-human and vehicle characteristics, road geometrics which suggested safety improvements, such as road geometric, infrastructure, and design standards. Therefore, different variables are studied individually or combined to reduce the collision rate and severity at road sections.

1.2 Road Safety Measures

For two decades, researchers have focused on road safety, and the current approaches toward improving road safety have been motivated by numerous management theories. Their view on the road safety system is based primarily on engineering. Mishra et al. [21] surveyed Delhi, India, particularly for two-wheelers collision collisions and found that the collision pattern, age, experience, average severity index of admitted patients is different from the reported in industrialized countries. Moreover, most collisions were recorded for two-wheelers (83%) on straight roads, whereas only 16 out of 87 were recorded as collisions with the car, bus, or truck. A large proportion of fatalities occurred at T-intersection (51%) and four-arm intersection (28%) and 79% at uncontrolled intersection studied by Bavishkar [5]. A study performed on the collision statistics of Vadodara city, Gujarat, India, by Sharma and Landge [35] proposes several safety measures for different causes of collisions. The measures included engineering, intersection improvement, enforcement, traffic education to road users, and safety enhancement at collision black spots. Pedestrians and two-wheeler riders are the vulnerable users, and 60% of victims died before reaching the hospital. Narayan et al. [24] and Chikkakrishna et al. [6] suggested intervention strategies to improve the safety of road users. Traffic segregation has been recommended to reduce collisions. Jain et al. [14] studied various parameters affecting injuries in 14 administrative wards in Pune, India, using univariate analysis. The maximum occurrence of the injuries was found in the age group of 15–30 and male drivers.

Roger [33] analyzed traffic collisions in four states for the mid-blocks of urban arterials in four states. The author found that the presence of medians results in a higher speed of vehicles. The absence of pedestrian and bicyclist facilities leads to a more significant number of collisions. The rate of collisions for two-wheelers increased from 18% (1978) to 39% (2013). The best solution is to include road safety issues in the school curriculum to reduce the severity of collisions and ensure that the vehicles follow the speed limits by Sarin et al. [34]. Epidemiological trends of the collisions arose that 84% occurs in urban areas and mainly 46.7% on National Highways represented by Rajaraman [31]. The auto-rickshaws have low fatal collisions compared to the other vehicle classes as their speed is limited to 40 km/h when carrying passengers.

1.3 Challenges of Road and Traffic Conditions in India

Different researchers recommend road safety measures for the various parameters, like road users, geometric design, driver behavior, and vehicular characteristics based on analyzing the critical areas of the road network. In India, drivers, bicyclists, and pedestrians are the most vulnerable to road collisions due to their risky behavior. Due to inadequate space on the roadside, the non-mortised vehicles, pedestrians,

Fig. 1.2 Examples of road safety challenges in India

(a) High risk pedestrians (b) Obscured sign board

(c) Improper curve design (d) Aggressive driving

and the bicyclist is the major challenge faced while designing and constructing the road features. The pedestrians are at high risk when no separate footpaths or lanes are provided and delineated (see Fig. 1.2a). Moreover, free left-turning movement at an intersection must be avoided to allow the movement of pedestrians during that duration of time.

The mandatory use of a helmet for two-wheelers motorists and seat belt compulsion for four-wheelers is made. However, only 60% of road users follow the compliance of helmets and 65% compliance the seat belt provision for the front seat. In one of the studies of Tamil Nadu in 2021, it is observed that the application of a seat belt for the rear seat was recorded to be zero percent. So to overcome the issues of not following the rules and regulations, active safety technologies like automatic braking, pedestrian detection, and seatbelt lock need to be implemented to reduce the severity of road collisions.

The speed control on the road is also a point of concern, as road design is done for the 85th percentile speed to reduce collisions. However, the speed of vehicles on the road is influenced by geometric design, surrounding land use, traffic flow, driving behavior, and the proportions of pedestrians, cyclists, buses, and curb side parking. Hence, to overcome the speed issues, speed humps, dead-end streets, and traffic islands are the few solutions depending on road and traffic characteristics. Improper location of signboards and signals is also a significant issue. The information and warning sign boards are essential for the safety of road users. However, contrary to the code provisions [13], the boards are placed inappropriately and not properly visible to the road users, as shown in Fig. 1.2b.

Road geometric design consistency is an emerging factor in highway designing. Recognizing and handling the inconsistency on the highway can significantly improve the safety performance on the road. The design of the inappropriate curve, the collision is illustrated in Fig. 1.2c. Considerable research has been conducted to explore the concept and identify the potential consistency measures. The improvement in the traffic circle at the intersection and roundabout need to be modified by modern practices. Moreover, for NMT (non-motorized transport), a separate lane is required to avoid the conflicts between the motorized and non-motorized vehicles. To implement the modified Road Safety Audit (RSA) for roads with a significant proportion of the collisions needs to be performed and evaluated. The RSA will identify the primary cause of the collision in the particular section and suggest safety measures.

Besides road and traffic characteristics, driver behavior plays a vital role in road safety. The gap acceptance can be defined as an essential factor in studying the relevant features related to capacity, delay, and road safety concerning driver's performance. Gap acceptance behavior to accept or reject the gap depends on the decision process of a driver while lane changing or during the maneuver. It generally depends on the critical gap and follow-up time. In contrast, the critical gap value is a changing parameter for each case which is predicted with the assistance of several models based on altered experimentations by researchers. Mohan and Bawa [22], studies the effect of inadequate headway between the vehicles and inappropriate judgment of the driver at the road section results in the collision. Road collisions are affected by the inappropriate movements of the vehicles between the vehicles and other road users or features.

The gap acceptance is a significant constraint to study the driver's behavior and analyze road safety. An incorrect estimate of the critical gap leads to wrong design decisions of the road elements. Moreover, Hurwitz [10] presented that 95% of the collisions occur due to the driver's error which is inadequate gaps that may be due to aggressiveness. The driver's gap acceptance depends on various factors like age, gender, type of vehicles driving, number of passengers, month determined by Elango et al. [8]. In contrast, the factors associated with site condition are the number of lanes, speed, traffic volume at the major and minor streets and type of traffic control device presented by Ashalata and Chandra [2]. Several researchers have worked upon the effects of drivers' behavior on collisions due to gap acceptance, which concluded that male and teen drivers have more aggressive behavior than females and adults. Therefore, when a driver makes a wrong gap decision, there is a strong probability of collision. Based on the available literature, it is also noted that, significant collisions occur due to the frequent lane changes that are part of the aggressive driver behavior on straight roads, as shown in Fig. 1.2d.

In this section, the problems confronted by the road and traffic characteristics from the engineering perspective are discussed, addressing the effect of speed, geometric design, driving behavior on the safety of the road. The difficult task is to implement a traffic management system after knowing the concerns of road collisions and fatalities. This system and its implementation challenges in India are discussed next.

1.4 Road Safety Management

1.4.1 RSM Methods and Process

Road Safety Management (RSM) is a multi-stage practice to ensure a comparatively safer road environment with the help of a well-structured process. Road Safety Management (RSM) includes the process of formulating the safety policy framework with effective implementation measures by defining clear objectives and responsibilities of every stakeholder to evaluate and modify the safer practices [23]. The organization for economic co-operation and development (OECD) [25] illustrates the RSM process to slice down the number of road collisions and related fatalities. With diverse organizational structures, various process cycles, the tool of analysis, and dynamic conclusions, researchers refer to RSM as a complex system [23, 27]. Based on many factors in the RSM analysis in practice, an organization's effectiveness is considered a prerequisite for effective RSM [39]. Furthermore, the effective RSM plans can support the adoptions by decision-makers for rectifying the safer road improvement practices, according to the International Transport Forum [12]. Therefore, good RSM plans must integrate the following key areas:

a. National laws and standard practices to perform RSM,
b. Network coverage for road sections,
c. Clearly defined roles and responsibilities for every stakeholder,
d. Advanced analysis tools to strengthen RSM, and
e. The advances to overcome the shortcomings of RSM.

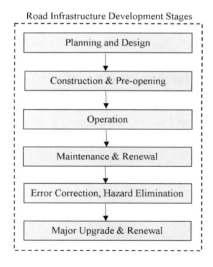

Fig. 1.3 Stages of road infrastructure project development

The road infrastructure development projects involve six significant stages, as shown in Fig. 1.3 [9]. With the integration with different project stages, the RSM includes the following ten methods [12]:

i. *Road Safety Impact Assessment.* The deliberate relative assessment is to be done at the preliminary stage to estimate the proposed project's road safety impact. The assessment also includes the rectification made to any existing road stretches. Therefore, the method plays a crucial role for RSM as it is applied before the project approval stage.

ii. *Efficiency Assessment Tools.* The spending of project budget costs must be dining in an optimized manner to ensure the project's financial viability. The Efficiency Assessment tools (e.g., sensitivity analysis) support the investment prioritization of the safest alternatives by analyzing the efficiency of specific investments.

iii. *Road Safety Audit.* The RSA is a well-defined checklist for ensuring safer road designs and management. The objective of RSA is to determine the lacking features of road infrastructure at every project stage (e.g., design stage) to ensure the implementation of the safest road environment features.

iv. *Network Operation.* This method transmits to timely managerial activities of the road networks to ensure the safety of every road user and the serviceability of road features.

v. *Road Infrastructure Safety Performance Indicators (SPI).* The performance indicators for road safety include road collision-based improvement measures and integrates the causes that lead to road collisions. The SPIs focus on identifying road safety hazards based on geometric design standards and infrastructural services. (e.g., error in curvature design).

vi. *Network Safety Ranking.* This method aims to identify, analyze and classify the road sections, specific places, or road networks based on their severity performance and collision costs.

vii. *Road Assessment Program (RAP).* The road assessment programs effectively collect road featured data sets and integrate the collected data sets for prioritizing the safety lacking. The road safety rectifications drives are formulated based on RAP performance reports.

viii. *Road Safety Inspection.* The accredited road safety inspector's team inspects the road network, severe road stretches, and collision blackspots. The systematic approach of RSI results as the road collision prevention tool with concluded detailed reports, which indicates the hazardous situation of study stretches with preventive approaches to tackle the hazards.

ix. *High-Risk Sites.* This method aims to determine the high collision risk roads sites that cause most of the collision fatalities for the last three-year.

x. *In-Depth Investigation.* This in-depth road safety investigation includes the most objective evidence of on-site road safety conditions (e.g., near collision-prone zones). The investigation concludes with the following information: (a) the principal reasons for a road collision, (b) collision victim's injury details and mechanism, and (c) rectification to prevent the specific collision.

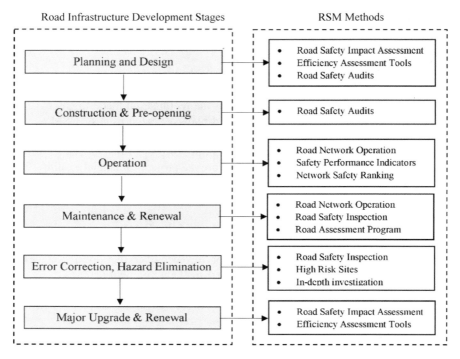

Fig. 1.4 Integrating RSM methods with road infrastructure development stages

Each of the RSM processes directly involves one or more project stages, as shown in Fig. 1.4. The proactive application of RSM demonstrated that most of the methods are applied at the normal operational stage and the project maintenance stage. In addition, the application of the RSM procedure also directly depends on the road classifications and geometric design rectifications. Every road has functional characteristics with specific traffic volume and vehicle composition. Based on this fact, the collision risk also affects the changing nature of road traffic characteristics. Therefore, the data and tools required for safety analysis tools keep changing with the evolving nature of those characteristics. The special requirements for a specific country must be considered while selecting the applicable RSM method, since driving standards, design standards, vehicle characteristics, and safety standards for every country are generally different. Table 1.3 elaborates on the factors affecting the selection of the RSM method.

Table 1.3 Applicability of various RSM methods

RSM method	Applicable road class	Range (section or network)
1. Road safety impact assessment	All road classes	– Road section
2. Efficiency assessment tools	All road classes	– Road section
3. Road safety audit	All road classes	– Rectified road infrastructure
4. Network operation	All road classes[a]	– Administrated road stretch – Share of the road network
5. Safety performance indicators	Rural road	– An intact road network
6. Network safety ranking	All road classes	– Administrated road stretch – Share of the road network
7. Road assessment programs	Rural road	– Road section – An intact road network
8. Road safety inspection	All road classes	– All road networks
9. High-risk sites	All road classes	– Administrated road stretch – Share of the road network
10. In-depth investigation	All road classes	– Investigation sites

[a] Challenging to apply in urban areas

1.4.2 Challenges of Indian Road Safety Management

As previously discussed, the collision scenarios in India along with the safety measures and challenges for road and traffic characteristics. However, the management of road safety is also of prime concern. The RSM challenges are grouped in the following categories: (a) management framework and policy challenges, (b) engineering design and standardization challenges, and (c) awareness challenges. The first category includes lack of management policy framework and law enforcement, lack of interdepartmental coordination, low inter-ministerial coordination, funding issues, and unreliable road collision data collection. The second category includes poor road design approach, lack of road safety devices, and lack of consultation with road safety experts. The last category includes lack of road safety knowledge among road users, lack of proper training and education, lack of proper communication channel, and prioritizing urban over rural areas.

Descriptions of these challenges are presented in Table 1.4. In addition, the challenges are graphically depicted in Fig. 1.5. For example, the policy implementation problems include the lack of continuity in government policies, inadequate human and material resources, all of which often lead to widening the implementation gap between the stated policy goals and the realization of such planned goals. Also, the active participation by stakeholders is expected for undergoing planning, creating benchmarks, setting up milestones and timelines for road safety activities for effective implementation of road safety map. There are several issues in road safety that demand coordinated efforts from multiple stakeholders. Close coordination among stakeholders shall pave the way for achieving goals followed by integrated monitoring of the activities. Interdepartmental coordination and operational monitoring shall be required at state, district and sub-district levels and a robust institutional mechanism. Road safety programs can be effectively and efficiently managed through more funding from national and local organizations, strong oversight by private–public boards, sound financial management, direct user charges, and regular technical and financial audits.

1.5 Opportunities for Improving Road Safety Management in India

To improvise the management system for road safety based on infrastructure, standard provision, policymaking, and enforcement of laws and regulation is discussed discretely in this section.

Developing Road Safety Manual for India

There is a need to develop a road safety manual for India. Numerous countries and organizations have developed such manuals, such as the Highway Safety Manual (HSM) by the American Association of State Highway and Transportation Officials United States [1] and the global Road Safety Manual by the Permanent International Association of Road Congresses [28]. For example, the HSM provides guidance for incorporating quantitative safety analysis to estimate highway safety performance in the planning and development of highway projects. The manual includes guidance related to human factors, roadway safety management process (e.g., network screening, site assessment, and project prioritization), and predictive methods for the analysis of infrastructure improvement project alternatives.

Improving Road Safety by Considering Worldwide Practices

The surface conditions for existing Indian roads are the most dangerous factor as compared to other collision-causing factors. So, the regular maintenance and patching work for the deteriorated road surface becomes essential to ensure safer vehicle

Table 1.4 Road safety management challenges in India

Safety challenge	Description
(A) Management framework and policy challenges	
1. Lack of integrated policy framework for road safety management and law enforcement	The apparent lack of integrated policy is noticed in the present Indian road safety scenario. Although the government is putting their efforts into devolving the enforcement and law policies to develop safer road culture, the integrated policy framework must be implemented for interdisciplinary straitening of Road safety management
2. Lack of interdepartmental coordination for promoting road safety	India is facing an alarming road collision scenario based on many factors. There must be perfect coordination between every stakeholder and department responsible for ensuring a safer road environment. Presently, India's situation highlights the lack of coordination to monitor the engineering standards and traffic law enforcement, strengthen the safer vehicle manufacturing standards, and ensure the quick post-collision medical response and other emergency care
3. Laid-back political will and low inter-ministerial coordination	The Central government and states governments are putting very few initiatives to promote the RSM within the country. Also, the lack of notable priority to enhance the present India RSM by engaging all the stakeholders is missing
4. Funding issues	Financial support for road safety awareness and training initiatives is rare from private bodies or the government. This situation also discourages technological changes and major transformations for road safety in India
5. Unreliable data collection systems for traffic collisions	The present collision recording process in India needs significant reformation. The fault collision data with lacking information cause collision analysis and reformation errors
(B) Engineering design and standardization challenges	
1. Poor road design approach	The design approach has minimal inputs from the road users. Some designs are more technically oriented and less user-friendly, causing hectic traffic conditions
2. Lack of road safety control devices	The Indian roads are noted with adequate safety furniture. However, the road design lacks mandatory signs, informative signs, collision barriers where required, and traffic calming devices

(continued)

Table 1.4 (continued)

Safety challenge	Description
3. Lack of consultation with road safety experts	Most road development projects do not have safety experts, or the safety experts are consulted on the last stage if administration by government officials only
(C) Awareness challenges	
1. Lack of road safety knowledge among road users	Most road users are not very much aware of safe driving practices and rules. Also, the road users are not able to interpret the meaning of every road sign and markings
2. Absence of proper training and education	There is a clear traffic rule and enforcement educational programs for traffic rules and enforcements for the general public. Also, there are no evident road safety awareness materials available in the regional languages of India
3. Lack of proper communication channels	The citizens are unaware of contacting which department to address their issues regarding road safety and the lacking road infrastructure. Also, the government stakeholders do not have a proper channel for interacting with road users and get aware of their inputs on a safer road environment
4. Prioritizing urban over rural areas	Most of the road safety rectification drives are urban area-centric. Often the rural collision-prone zones are neglected, and hence it becomes difficult to ensure safety in a rural area

movements [3]. The road administration bodies and consultants must adopt modern delivery tools and promote the rectification programs for safer road infrastructure development.

Developing Minimum Standards for Every Road Class

As per the investigations provided by Road Assessment Program (RAP), advancing existing road infrastructures in the direction of a 3-star rating may reduce collision fatalities, according to the International Road Assessment Program [11]. Many developed countries like Sweden and Netherland aims to improve their existing road networks to achieve minimum 3-star rating. The implementation of similar practices will ensure the overall safety management of Indian roads.

Evaluating Rectification Measures with Detailed Impact Assessment

The timely and continuous Road Safety Management may reduce the project's overall cost at its best [12]. The complex road safety scenario of India demands the inclusion of continuous evaluation of road safety rectifications. Also, it needs to implement

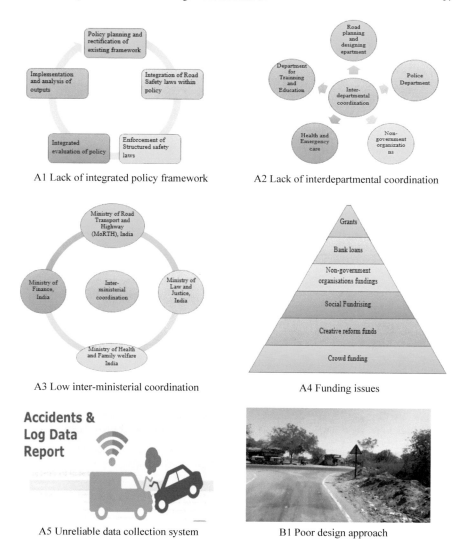

Fig. 1.5 Graphical depiction of road safety management challenges in India

the benefit analysis tools to assess the detailed impact. The impact assessment also helps the road safety authorities to deal with the limited funds.

Formulating Well-Defined RSM Structure for Every Stage of Road Projects

The government has to form a hierarchical structure with well-defined work plans for every road safety stakeholder. Although the engineering design significantly contributes to ensuring road users' safety, other factors like enforcement, public

B2 Lack of road safety control devices

B3 Lack of consultation with safety experts

C1 Lack of road safety awareness campaign

C2 Lack of training and education

C3 Lack of proper communication channel

C4 Prioritizing urban over rural area

Fig. 1.5 (continued)

awareness, and emergency health care also play crucial roles in developing effective RSM plans. Based on this fact, every road project must be enforced with a proper RSM plan which integrates all 4Es (Engineering, Enforcement, and Education, Emergency care) to ensure road safety.

Making Road Safety Management Legally Mandatory

At present, the road safety auditing process and road safety management plans are not mandatory in India. Therefore, a road safety audit is done in unavoidable situations. The judiciary system of India must analyze the alarming road safety scenario and make RSM mandatory for government and private road development bodies.

Improving Collision Data Collection and Management Systems

The accurate road collision data collection impacts the effectiveness of RSM. Therefore, India must improve data collection approaches of traffic collisions to ensure accurate road collisions by implementing real-time data collection systems. Furthermore, the data collection and management system for collision indicators must consider the collision influencing parameters, such as deficiency of designing, condition of vehicles involved in a collision, driver's condition, and condition of safety infrastructure.

Establishing Training Programs for all Road Users

The drivers and most road users are avoiding getting involved in safety education drives. Therefore, the government and private bodies must conduct road safety awareness and educational drives in local languages to ensure trainees' highest level of integration. Also, periodic safe driving skill training programs are to be initiated for timely skill checks and improvement in vehicle driving.

Developing Specific Safety Approach for Rural Areas

Most of the deliberations and remedies should ensure that road safety is more urban-centric. The safety scenario for rural India requires special attention with tailored-made rectifications. The rural road safety committees have to structure within every Indian state, and regular safety visits need to be done for collision-prone rural stretches. Also, rural road users will need to be aware of the essential practices to ensure their safety at most.

Mobilizing Available Resources

India road collision scenario demands funds rising with more requirements of resources. The state governments and central governments need to allocate specific budgetary allocations for road safety improvement. Non-government bodies like safe NGOs, automobile firms, insurance companies will need to be invited for safety fundraising initiatives. Lastly, the disbursement of the funds must be supervised by legal government bodies with timely checks.

Promote Political Backing and Spirit

The national-level involvement and rectifications require enough backing from political parties and political representatives. The road safety awareness messages and initiatives must be included within public gatherings, awareness notifications, and national-level political agendas. Also, the authorized ministers must structure, monitor, and implement the national road safety policies and agendas with a great spirit and involvement.

Figure 1.6 represents the integration between the road safety challenges and opportunities for challenges and management for Indian road conditions. The management challenges are linked with the road safety opportunities to overcome the challenges using the various check for the opportunities as depicted in the figure above.

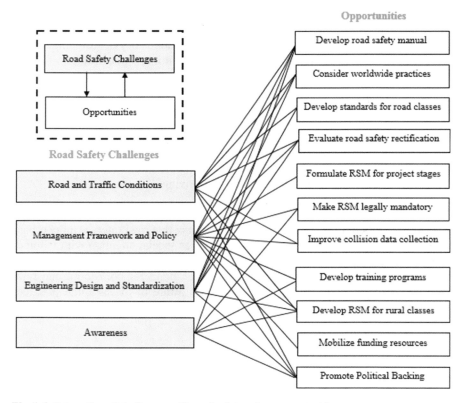

Fig. 1.6 Integration of challenges with road safety reform opportunities

1.6 Conclusions

Traffic collisions have increased for few decades, and hence road safety is a challenge to reduce the number of collisions, injuries, and fatalities. It is also an important aspect to reduce the risk of road collisions. Considerable research has been conducted to study various aspects, including collision scenarios, collision modification factors, risk models, and safety measures. Road collisions have various causes depending on the parameters of road users and geometric and traffic characteristics. Based on this study, the following comments are offered:

1. The major causes of the collisions at a particular section can suggest the improvements related to the road design. However, the road user's behavior is a concerning issue for road safety. The traffic rules and regulations are enforced for the safety of vehicle users. Regarding safety, the major challenge in India is the management of safety measures. The present Indian road collision scenarios demand immediate actions to ensure the safety of all road users.

2. In the last five years, collision trends suggest that the number of road collisions in India is decreasing, but the fatalities caused by road collisions are increasing. The increased percentage share of road fatalities is the central concerning area for the Indian administration. The higher usage of motorized vehicles, specifically more two-wheeler occupancy, unbalances the safety infrastructure for non-motorized usages. There is a need for adequately integrated efforts for engineering measures, law enforcement, safety education, and emergency care adversely affecting the road safety in India. Additionally, road users need a systematic training and awareness program to brush up on safer road practices. Therefore, the road safety management procedure and strategic improvements based on the management structure is the "perfect package for Road safety" in India.
3. A state-of-the-art road safety management plans would help policymakers and officials to establish an efficient safety management structure. The road safety management will enable the integration of different Indian stakeholders working individually for road safety, like the transport design engineers, safety auditors, policymakers, enforcement agencies, police departments, and health professionals. Also, accurate safety management will provide a sense of responsibility with working freedom to every Indian safety professional. This will allow them to work individually with a sense of responsibility to respect interdepartmental opinions and ideas.
4. India may effectively handle the alarming road safety situation with the presented context by prioritizing good road safety management policies and frameworks. The clear progression for the road safety management scheme is considered as a strong base to improve road safety in India, and this vital field needs a detailed study.

References

1. American Association of State Highway and Transportation Officials United States. (2010). *Road safety manual*. AASHTO.
2. Ashalata, R., & Chandra, S. (2011). Critical gap through clearing behavior of drivers at unsignallised intersections. *KSCE Journal of Civil Engineering, 15*, 1427–1434.
3. Bank Asian Development. (2012). *Performance-based routine maintenance of rural roads by maintenance groups. Guide for Communications Bureaus.* Asian Development Bank. https://www.adb.org/sites/default/files/publication/30090/performance-based-routine-maintenance-rural-roads-guide.pdf
4. Bashar, A., Ghuzlan, K., & Hasan, H. (2013). Traffic accidents trends and characteristics in Jordan. *International Journal of Civil Environmental Engineering, 13*, 9–16.
5. Baviskar, S. (1998). Analysis of road accidents on national highways in Nashik district. *Indian Highways, 26*, 37–48.
6. Chikkakrishna, N. K., Parida, M., & Jain, S. S. (2013). Crash prediction for multilane highway stretch in India. *Proceeding of Eastern Asia Society for Transportation Studies, 9*.
7. Dinu, R., & Veeraragavan, A. (2011). Random parameter models for accident prediction on two-lane undivided highways in India. *Journal of Safety Research, 42*, 39–42.

8. Elango, S., Ramya, A., Renita, A., Ramana, M., Revathy, S., & Rajajeyakumar, M. (2018). An analysis of road traffic injuries in India from 2013 to 2016: A review article. *Journal of Community Medicine & Health Education, 8*.
9. Elvik, R. (2010). Assessment and applicability of road safety management evaluation tools: Current practice and state-of-the-art in Europe.
10. Hurwitz, D. (2011). Connecting gap acceptance behavior with crash experience. *Road Safety Simulation*, 1–18.
11. International Road Assessment Program. (2014). *IRAP star rating and investment plan coding manual*.
12. International Transport Forum. (2007). *Road infrastructure safety management road infrastructure safety management*.
13. IRC:93-1985. (1996). *Guidelines on design and installation of road traffic signals*. Indian Road Congress.
14. Jain, S., Singh, P., & Parida, M. (2011). *Road safety audit for four-lane national highways*. Indianapolis.
15. Jha, N., Srinivasa, D., Roy, G., & Jagdish, S. (2003). Injury pattern among road traffic accident cases: A study from South India. *Indian Journal Community Med, 28*, 84–90.
16. Kadiyali, L., Gopalaswami, T., Lakshmikanthan, P., Pathak, U., & Sood, A. (1983). *Effect of road characteristics on accident rates on rural highways in India*. Highway Research Bulletin.
17. Kanchan, T., Kulkarni, V., Bakkannavar, M., Kumar, N., & Unnikrishnan, B. (2012). Analysis of fatal road traffic accidents in a coastal township of South India. *Journal of Forensic and Legal Medicine, 19*, 448–451.
18. Kar, S., Das, S., Tiwari, A., & Pharveen, I. (2014). *Pattern of road traffic accidents in Bhubaneswar, Odisha*. Clinical Epidemiology and Global Health
19. Ministry of Road Transport and Highway. (2018). *Road accidents in India; New Delhi*.
20. Mirkazemi, R., & Kar, A. (2014). A population-based study on road traffic injuries in Pune City, India. *Traffic Injury Prevention, 159*, 379–385.
21. Mishra, K., Banerji, K., & Mohan, D. (1984). Two-wheeler injuries in Delhi, India: A study of crash victims hospitalized in a neuro-surgery ward. *Accident Analysis & Prevention, 16*, 407–416.
22. Mohan, D., & Bawa, P. (1985). An analysis of road traffic fatalities in Delhi, India. *Accident Analysis & Prevention, 17*, 33–45.
23. Muhlrad, N. (2009). *Road safety management systems, a comprehensive diagnosis method adaptable to low and middle income countries. Synthèse INRETS*.
24. Narayan, S., Balakumar, S., Kumar, S., Bhuvanesh, M., Hassan, A., Rajaraman, R., & Padmanaban, J. (2011). *Characteristics of fatal road traffic accidents on Indian highways*.
25. Organization for Economic Co-operation and Development. (2002). *Road safety: What's the vision?* http://www.internationaltransportforum.org/Pub/pdf/02SafetyOnRoads.pdf
26. Pandey, G., Mohan, D., & Rao, R. (2015). Why do three-wheelers carrying school children suffer very low fatal crashes? *IATSS Research, 38*, 130–134.
27. Papadimitriou, E., & Yannis, G. (2013). Is road safety management linked to road safety performance? *Accident Analysis and Prevention, 59*, 593–603. https://doi.org/10.1016/j.aap.2013.07.015
28. Permanent International Association of Road Congresses. (2019). *Road safety manual*. PIARC. https://www.piarc.org/en/
29. Prajapati, P., & Tiwari, G. (2013). Evaluating safety of urban arterial roads of medium sized Indian city. In *Proceedings of the Eastern Asia society for transportation studies*.
30. Raichur, M., Panwala, C., Chawda, S., & Shah, J. (1993). Road safety in and around Vadodara City. *Indian Highways, 21*, 7–15.
31. Rajaraman, R. (2009). *Analysis of road traffic accidents on NH45, Kanchipuram District (Tamil Nadu, India) Seoul, Korea*.
32. Ratanavaraha, V., & Suangka, S. (2017). Impacts of accident severity factors and loss values of crashes on expressways in Thailand. *IATSS Research, 37*, 130–136.
33. Roger, L. (2012). IRAP India the four states road safety report.

34. Sarin, S., Chand, F., & Rao, V. (2005). Minor improvements of highways for better road safety. *Indian Highways, 33*, 15–54.
35. Sharma, A., & Landge, V. (2012). Pedestrian accident prediction model for the rural road. *International Journal of Science and Advanced Technology, 2*, 66–73.
36. Singh, D., Singh, S., Kumaran, M., & Goel, S. (2015). Epidemiology of road traffic accident deaths in children in Chandigarh zone of North-West India. *Egyptian Journal of Forensic Sciences*.
37. Singh, S. (2017). Road traffic accidents in India: Issues and challenges. *Transport Research Procedia, 25*, 4708–4719.
38. Singh, S., & Mishra, A. (2001). Road accident analysis: A case study of Patna city. *Urban Transport Journal, 2*, 60–75.
39. Varhelyi, A. (2016). Road safety management—The need for a systematic approach. *The Open Transportation Journal, 10*(1), 137–155. https://doi.org/10.2174/1874447801610010137

Chapter 2
Port Development: History, Present and Future Challenges

Abhijit Singh

Abstract Over the years, the role of ports has undergone a significant change. From being only cargo handling points, ports first evolved to take care of other items of the supply chain like warehousing, bagging, value addition to cargo before and after shipment, etc., and later other items of the logistics ecosystem like multimodal facilitation, cargo movement tracking, maximization of fleet usage, etc. Managing this vital mode of transportation is very much important. Infrastructure planning and management plays a crucial role towards a healthy outlook for the ports sector. Unfortunately, Indian ports are far from satisfactory level in terms of their efficiency, time required in custom clearance and many other. Infrastructures of Indian ports are in a very poor stage. In this background, the proposed chapter intends to understand the various bottlenecks and challenges in Port Infrastructure Development in India. This chapter will also intend to identify various stages of port development and future courses along with lessons learnt from the past and the challenges expected in future with special reference to India for sustainable port infrastructure development to increase international trade of India.

Keywords Port · Shipping · Port management · Port development

2.1 Introduction

The role of ports has evolved significantly over the years. From being only cargo handling points, ports first evolved to take care of other items of the supply chain like warehousing, bagging, value addition to cargo before and after shipment, etc., and later even other items of the logistics ecosystem like multimodal facilitation, cargo movement tracking, maximization of fleet usage, etc. Today, as an active stakeholder in 'Ease of Doing Business', ports have made the interaction with port users/trade and industry much easier, paperless and fast.

A. Singh (✉)
Executive Director, IPA (An Apex Body of Major Ports Under Administrative Control of Ministry of Ports, Shipping and Waterways, Govt. of India), New Delhi, India

© The Author(s), under exclusive license to Springer Nature Singapore Pte Ltd. 2022
P. Jadhav and R. N. Choudhury (eds.), *Infrastructure Planning and Management in India*, Studies in Infrastructure and Control, https://doi.org/10.1007/978-981-16-8837-9_2

Scope of port development is no more confined to its core service offerings. Over the last 50 years, a number of prominent port-centred industrial clusters have emerged. The port authorities world over are facing more and more challenges with the globalization of terminal operations. During the 1990s, terminal operators and major shipping companies began to band together to invest in and control terminals all over the world.

This globalization trend has primarily impacted containerized terminal operations. These days, the major global container trade is increasingly controlled by a small number of top-tier carrier alliances and independent terminal operators [1].

2.2 Port Development History

The maritime activity of cargo and mass movement has always been the most cost-effective, suitable and practical mode of transportation. This is why, since at least 6000 BC, the world has been developing ports. Egyptian history, which dates back to 4000 BC, provides evidence of using waterway transportation as a mode for transportation of stones and other raw material for the construction of the Great Pyramids. It is believed that earliest specialized freight handling facilities that were built in the human history for maritime trade were developed for the construction of the pyramids. Ancient Romans too had a formidable navy and merchant ships and harbours from where ships travelled to North Africa and other European countries.

The Indian maritime history is believed to have begun during the 3rd millennium BC when settlers of the Indus Valley Civilization started maritime trade and relations with Mesopotamia. The Roman historians mention about extensive trade by Romans with India after the invasion of Egypt. As per their accounts, by the time of Augustus reign, more than 100 ships were sailing every year from Myos Hormos, a red sea port built around third century BC, to India. With increase in trade between India and the Greco-Roman civilization, spices became the principal export from India to the Western world, overtaking silk and other commodities. Even long after the fall of the Roman Empire, Indians continued to reside in Alexandria while Christian and Jew immigrants from Rome had their presence in India. This resulted in Romans losing the Red Sea ports, which hitherto were used by the Greco-Roman empires to safeguard the trade with India since the Ptolemaic dynasty. The trade relationship between India and Southeast Asia proved critical to the Arabian and Persian merchants between the seventh and eighth century AD [2].

2.3 Evolution of Modern Ports

Over the past two centuries, the port sector has changed radically, where during nineteenth and first half of twentieth century, ports were considered instruments of state or colonial power as well as a strategy for market control. There was limited competition and costs associated with the ports meagre relative to the overall cost of transportation. Thus, there were no incentives to improve port efficiency and service levels. However, with rapid commercial expansion and industrialization, the seaports were transforming in the way they handled cargo and shipping, as well as their interaction with urban growth. By the early nineteenth century, with growing business, Antwerp, London, Liverpool were among the European ports who had already invested in docks; however, their design could not anticipate the impact of steam engines. This was incorporated in subsequent stages of port construction during and after mid-century to meet the demands of bigger vessels, faster turn-around time and better facilities for bulk cargo. By late nineteenth century, progression from sail to steam, emergence of new trades and trade routes and evolution of liner business posed a significant challenge as well as opportunities to the seaport industry. Not all nineteenth-century sea port development was a result of strengthening of the established business; new settlement areas also gave rise to many new ports, and connectivity to the hinterland was a principal factor that influenced the development of numerous ports.

After the First World War, the period of expansion came to an end, and due to the inter-war trade depression, there was less demand for the facilities of the port and more emphasis on maintaining profitability. Nevertheless, investment in construction of new wharf, quay and port-basin continued but with a slower pace and most of the new projects inclined towards terminals for petroleum and oil, which often times shifted port activity from their erstwhile locations. The effect of trade revival after the Second World War in Europe was initially suppressed by the scale of wartime devastation, but by the mid-1950s, most Northern European sea ports were handling almost the same quantity of cargo as they had before the World War II.

In other parts of the world, decolonization led to more aspiring proposals for port development and pressure on shipping increased along with the surge in international trade. Oil tankers as well as the ships in the dry bulk trade grew in size to leverage economies of scale thus giving rise to demand for deep berth terminals. In case of general cargo traffic, it was not the quantity and size of cargo but its packaging which challenged the innovation and creativity of port managers in the 1960s and 1970s leading to assessment of unitization and its implication on port efficiency and productivity. Furthermore, new and innovative business models and advancements in maritime technology presented multitude of challenges to the effective planning and administration of seaport assets and infrastructure [2].

By late twentieth century, number of maritime economies saw a trend of inclination towards port privatization, complete or partial. The involvement of private enterprise manifested a change in port efficiency, productivity and dynamism, thus advancing the port sector in to twenty-first century which further brought new and radical transformation in port operations [3]. Increasingly vigorous global competition compelled ports, as well as other stakeholders in the worldwide logistics chain, to change the way they conducted business going forward. Innovative systems and induction of new technologies changed the requirements for port infrastructure radically and increased the degree of specialization requirement in port operations, thereby raising the financial stakes in port projects and the need for a highly skilled and trained workforce. Also, the late twentieth century's shipping revolution in cargo handling transformed the terminal-related labour process itself. Modernized ports started characterizing themselves with high degree of automation, digitization, sophisticated communications technologies, high skilled workforce and all-year 24 × 7 terminal operations [3].

2.4 Port Governance Mechanism

Ports over the time have undergone a rapid change. A change in economic and business environment due to globalization of production and distribution centres, technological advancements, transformation of cargo transportation business models, etc., concluded a long era of mostly public (government) controlled port governance models across the world. Port reforms started gaining traction in late 1980s and early 1990s with desire to have more than before private participation in the sector.

The core idea was to transfer the port management and operational responsibilities to better commercially driven private entities, and port assets are still owned by the government or government-controlled port authorities along with regulatory, supervisory and monitoring role. However, in some cases like the UK, government shifted towards complete privatization with the out-right sale, including port land and regulatory functions, of port assets. Corporatization of state-owned ports has also been a leading governance model adopted by many countries. In certain nations, a sole port authority is in charge of all important ports, e.g., South Africa's Transnet National Ports Authority. The 2008 global economic crisis further gave the impetus to port reforms, mainly the transition towards the complete landlord model, with the explicit goal of boosting efficiency, increasing cargo traffic volumes and improving profitability of ports [4].

In India, public–private partnership (PPP) mode of development was introduced in the Port Sector in mid-1990s along with other infrastructure sectors. Since then, PPP mode has become the preferred mode of infrastructure development in major ports. Non-major ports, which are either private or on a long-term lease to private developers with complete autonomy, have also proliferated in past two decades. With the new Major Port Authority Act 2021, major ports are set to get more autonomy, thereby facilitating them transforming towards a complete landlord model. Kamarajar Port Limited (Ennore) near Chennai is a successful experiment towards establishing a corporatized major port in India. This could set a trend for the development of any new Greenfield Major Port in India to be established as corporate entity. Further, a new Indian Ports Bill is on anvil, replacing the current Indian Port Act of 1908, which would usher in a mechanism for structured and sustainable growth of overall Indian port sector.

2.5 Use of Concessions in Public Private Partnership (PPP)

As per the World Bank Group definition 'A Concession is an arrangement whereby a private party (concessionaire) leases assets from an authorized public entity (grantor) for a defined period and has responsibility for financing specified new fixed investments during the period and for providing specified services associated with the assets; in return, the concessionaire receives specified revenues from the operation of the assets; the assets revert to the public sector at expiration of the contract'.

Governance model based on concessions to private players of rights to operate port assets and deliver port-related services, and port authorities acting as landlords and retaining port ownership, is the modus operandi in vogue worldwide. In order to effectively engage with the experienced private operators, it requires port authorities or government administrations to assimilate the legal, commercial and institutional skills required to achieve balanced and lucrative arrangements. Even after that it is sometimes hard for authorities to deal with partners whose short-term objectives may be in conflict with public authorities' long-term policy goals and objectives. Landlord port model requires a lot of engagement on part of the government in putting together the various concessions of port and related activities, as well as redesigning and restructuring of the port organization.

In order to mitigate stakeholders concerns and ensure uniformity and consistency in the design of PPP contracts, in order to send a clear message to the market, many countries have adopted Model Concession Agreements (MCAs) as a standard procurement framework, clearly indicating what the standard PPP procurement process would likely be. This gives a clear sign to prospective private sector partners and is likely to reduce the disputes about the award decision. This also saves time and money of the prospective bidders by lessening the time required to comprehend each project contract and eliminates project-by-project negotiations as all parties have an understanding of what is acceptable. Moreover, it enables the authorities to concentrate more on developing and refining existing processes and legal documentation, rather than drafting concession agreements or contracts from scratch.

First Model Concession Agreement for Major Ports in India was approved by the Government of India on 3 January 2008. Earlier, projects were awarded by government based on project specific agreements and standard agreement developed by IDFC. MCA 2008 was approved with an objective to ensure uniformity in the concession agreements, for the Build, Operate and Transfer (BOT) projects to be developed under Public–Private Partnership (PPP) mode, between the major ports and the selected bidders. Further, MCA was revised and approved by the government in March 2018 with aim to make port projects more investor-friendly and make the investment climate in the sector more attractive. Lately a new Model Concession Agreement – 2021 (MCA 2021) formulated, with aim to facilitate greater private sector participation in PPP projects in major ports and ensure equitable and reasonable risk allocation, more transparency and promote competition. This will also bring more confidence of developers, investors and lenders and other stakeholders in the Ports sector and catalyze the investment in the sector.

2.6 Changing Role of Ports in Twenty-first Century

Ports are no more merely a node in a transport or supply chain network. Ports in twenty-first century have become the clusters of economic activity. Though, sea ports have always been a critical link in the logistics chain, but with production and consumption centres becoming globalized, it has given ports a unique opportunity to add value to the overall supply chain. Changing international logistics business model and trade routes, environment regulations, technological advancements, and industry consolidation are going to change the way ports will operate in future. The supply chain design has evolved rapidly over the years, because of constantly changing or developing customer demands or requirements concerning price, quality, flexibility, delivery, reliability and service [5].

With multinational manufacturing and logistics companies moving towards sophisticated supply chain networks and systems, the ports will have to position themselves as attractive partners. Port's competitiveness and efficiency plays important role in choosing a location for a factory or distribution centre and often determine the competitiveness of the local producer with other producers to compete globally or regionally. It will be up to the ports to meet their customer expectations and support them in them in gaining a competitive advantage by offering low-cost and highly efficient port services. To remove bottlenecks in the maritime supply chain effective coordination, communication and collaboration among stakeholders would be the key driver to success.

2.7 Digital Transformation of Ports

Like other sectors of economy, maritime industry has also been impacted by the new age digital technologies. Each generation of ports over the history of port development came with a new function and focus. With increased demand for digital integration, leading ports are focusing on transforming their function more as a service provider, not necessarily as physical services like towage and terminal operations, but more as a platform and data service provider. This means digitization of port activities and processes where new services would either replace or reinforce traditional port services. Advancing towards the next generation of smart ports, which makes full use of IoT networks and big data solutions, means that a port must be able to recognize and take advantage of new business models within this emerging and large scale ecosystem. A lot of the data generated could act as a building block for new, innovative products and services. Port insight is a great example of this innovation in the Netherlands, and Port of Rotterdam is developing Internet of Things solutions, to the benefit of the whole port community [6, 7].

Building new models would require integration between the supply and demand side from the transportation and logistics sector, assimilating not only logistics companies and suppliers and distributors but also their customers such as Original

Equipment Manufacturers (OEMs). Imagine a global network of ports where the ultimate consignee of cargo can track his consignments on real-time basis right from the point of source to his warehouse; where all documents are in standardized digital format and no manual intervention is required for their scrutiny and approvals; where waiting time at any touch point in overall supply chain network is minimal or negligible; where variability in deliveries is so negligible that manufacturers receive their orders just-in-time removing the need to keep any inventories; where all stakeholders can find relevant information and request for processing of documents or transactions on a single digital platform; where business operations turn more pro-active rather than reactive; all this and much more are possible with advance digital technologies and innovations. Thus, port digitization has capability to increase efficiencies of port operations so exponentially that overall system can run with much fewer resources thus helping ports becoming highly sustainable [8].

Ports are increasingly implementing innovations in the entire value chain with a wide range of used technologies. Globally most of the ports have moved to state of the art Terminal Operating Systems, Port Community Systems and ERP tools. Advanced and under development technologies such as Hyperloop, Automated Guided Vehicles, 3D printers are under pilot phases internationally and might become mainstream and relevant in time to come. 3D scanners and computer-enabled vision, which helps in collecting information and data from various visual inputs, have the potential to make stowage operations extremely efficient. Virtual and augmented reality technologies are providing unprecedented insights in to the business processes. Drones deployed at ports can undertake various tasks, such as equipment inspection, inventory monitoring and mislaid goods detection, and the surveillance and detection of trespassing, etc. Drones could also survey harbours and provide the incoming and outgoing vessels with a greater degree of precision to assist in vessel berthing and manoeuvring.

The port of Rotterdam will soon have its own 'Additive Manufacturing FieldLab' with 3D metal printers. In 2018, One DP World Cargospeed, an international brand for hyperloop-enabled freight transportation, was launched by DP World, Dubai's trade facilitator and the world's third biggest port authority, to facilitate the transportation of unitized cargo in a timely, cost-effective and environmentally friendly manner at the speed of flight and closer to the cost of trucking. TradeLens, an open and equitable technological solution developed by IBM and Maersk, based on blockchain technology and supported by major industry player, including more than 20 port operators and terminals around the globe including PSA Singapore, Modern Terminals in Hong Kong, Port of Rotterdam, Valencia Port, Port of Bilbao, and terminal operators Holt Logistics in the Port of Philadelphia.

Digital twins are among another logistics technology trends which industry is looking forward to. With digital twins, digital model of any physical asset or object can be created in such a way that allows users to interact with a digital representation of a real thing or its segment just like we would engage with their physical equivalent. Using technologies in ML and AI can help ports make more sense of the data and get better insights for improving efficiencies and driving down the cost. The blockchain technology and smart contracts can resolve the issue of trust among various stakeholders and make information sharing and transactions more secure and fast. A few experimental projects and small-scale activities are being carried out by some firms to this end. For example, CargoX is one such startup that has been working dedicatedly on bringing blockchain to the transport and logistics business using the public Ethereum network to securely authenticate document transactions. Likewise, various port authorities in Northern Europe have also been working on developing their own blockchain management platforms.

2.8 Port Community Systems

In the past few years, the world's leading ports have been investing in their digital ecosystems, in particular in creating Port Community Systems (PCS) to enhance their connectivity. Port Community Systems arrange a single entry platform of data in a port for business-to-government messages and the business-to-business messages thus boosting the efficiency by providing digital, real-time information to all stakeholders in the port. Most of the world's notable ports have PCS systems in operation.

Today's systems also need to focus more on the evolving dimension of worldwide connected ports and logistical chains. It is also vital to look ahead to new developments like the need for global connectivity, which goes far beyond the scope of connecting the local community. Ports will have to reach out much further into the hinterland and be connected to other seaports, thus making a global network of smart ports. To do so, requires the guts and willingness to work together, trust one another and be open and transparent with one another. Even competitors will be compelled to exchange information in order to provide more visibility to the end consumer in to the world of maritime freight, where at present there is minimal transparency. By creating a platform to port community members that are part of the port logistics chain, these community members will be able to optimize their processes. Through the sharing of data, port developers can remain in control of their own core tasks. This will strengthen the competitive position of the port and guarantee the safe and efficient handling of processes in the port.

In India too, in order to improve the ease of doing business and facilitate trade, government has undertaken a number of initiatives during the last few years to improve the flow of trade through seaports. One of the key initiatives is Port Community System (PCS 1x), a platform to exchange trade related documents electronically. A centralized web-based Port Community System (PCS) is functional across all major ports which allows seamless data exchange among the various stakeholders through a common platform. (PCS) is an initiative by Indian Ports Association (IPA) at the instance of Ministry of Port, Shipping and Waterways envisaged to integrate electronic information sharing and serve as a secure central hub for Indian ports and other stakeholders such as Shipping Lines, Agents, Banks, Stevedores, Regulatory Agencies, Surveyors, Customs House agents, Importers, Exporters, Container

Freight Stations, Railways, CONCOR etc. through a shared gateway. The implementation of an 'Open Platform' in PCS 1 × facilitates various time-tested solutions to connect with PCS and provide an unparalleled ability to enhance services to various stakeholders. The 'Latch On' feature is a unique concept built in and delivered with PCS 1x. The Latch On feature facilitates the trade in providing the required features that cannot be directly embedded into any Single Window Platform even though the features/ functionalities are required for seamless data and document exchange. Latch-ons achieve this without duplicating the effort.

2.9 Conclusion and Way Forward

Maritime transport is the cheaper and cleaner mode of transporting cargo from one place to another. Maritime history is as old as the history of modern human civilization. It has though transformed drastically over the last two centuries. A change in economic and business environment has put an end to a long period of mostly state—controlled port governance models and private participation in port development has become a norm. A robust port governance mechanism that is pro-active to dynamic and changing global maritime supply chain would determine the success and survival of the ports in future. A greater private participation that would bring in innovative technological solutions would be the way forward for the port authorities to overcome the challenges in future.

Ports are increasingly implementing innovations and are supplementing their role as value-adders in the overall supply chain. Digital technologies have developed rapidly over the years and are continuing to do so. Many global ports have already started harnessing the benefits of these technologies. Today leading global ports are collaborating to create a worldwide network of smart ports. To achieve this, it requires the courage and willingness to cooperate, trust and be transparent to each other. The future of the global ports and shipping industry may still look uncertain, but change in trade routes, competitive position of ports, ecosystems, and cargo distribution would decide the survival.

References

1. Juhel Marc, H. (2017). Container terminal concession guidelines. Sub-Saharan Africa transport policy working paper; No. 107. World Bank (© World Bank, License: CC BY 3.0 IGO). https://openknowledge.worldbank.org/handle/10986/28309
2. Sorgenfrei, J. (2018). Port business: Second edition (Revised, Enlarged ed.). DelG Press; World Bank. (2021, February). *The evolution of ports in a competitive world*. Port Reform Toolkit Second Edition. https://ppiaf.org/sites/ppiaf.org/files/documents/toolkits/Portoolkit/Toolkit/index.html
3. Notteboom, T., Pallis, A., & Rodrigue, J. (2022). *Port economics, management and policy* (1st ed.). Routledge.

4. Brooks, M. R., & Pallis, A. A. (2012). Port governance. In W. K. Talley (Ed.), *The Blackwell companion to maritime economics* (pp. 491–516). Wiley-Blackwell.
5. Palmer, S. (1999). Current port trends in an historical perspective. *Journal for Maritime Research, 1*(1), 99–111. https://doi.org/10.1080/21533369.1999.9668302
6. Buck. W., Gardeitchik, J., & Deij, A. (2019). Move forward: Step by step towards a digital port [white paper]. Port of Rotterdam. https://www.britishports.org.uk/system/files/documents/smart_port_papers.pdf
7. Molavi, A., Lim, G. J., & Race, B. (2019). A framework for building a smart port and smart port index. *International Journal of Sustainable Transportation, 14*(9), 686–700. https://doi.org/10.1080/15568318.2019.1610919
8. Pannekoek, M. (2019). Move forward: Move forward: Data as a fuel for digital port [white paper]. Port of Rotterdam. https://rotterdam-port.accounts02.wp-magazines.com/wp-content/uploads/sites/1684/2019/12/DBS_WP2_Data_as_fuel_11-03-19.pdf

Chapter 3
Emerging Scope of Airport Infrastructure: Case of India

Manish Yadav and Tarun Dhingra

Abstract Since 2014, India's commercial airline sector has augmented to become one of the country's fastest-growing enterprises. India has surpassed the UK to become the world's third-biggest regional airline marketplace, with the UK likely to fall behind in 2024. The Indian government has been trying to increase the number of airport terminals in order to meet the growing needs. India had 103 operating terminals as of March 2019, intending to build up that number to 190–200 by FY40. It is seen as a strategy to encourage commercial growth, with transportation infrastructure being required for the nation's economic sustainability and growth. Although infrastructure does not immediately culminate in development, it is seen as a necessity for any socio-economic and sustainable growth. To institute the scope, the chapter enlists the importance of airport infrastructure in the Indian economy enrooted from operational, planning and management outlook. This chapter will describe the aspect of operations, including airport classification, i.e. Kerbside, Terminal and Airside. This chapter attempts to pen down the fundamentals of airport infrastructure planning, including capacity planning, master planning, facility and layout planning. The chapter will also illustrate the management perspective of airport infrastructure and evaluate commercial revenues, including aeronautical and non-aeronautical revenues with financial sustainability. Finally, the chapter will address the role of the Airport Economic Regulatory Authority and Ministry of Civil Aviation in airport pricing mechanism, Multi-Year Tariff Planning (MYTP) and overall revenue approaches, especially during public private partnerships and impacts on airport competition, growths and sustainability contributing to the Indian economy and its sustainability through the aviation sector paving the way forward for establishment of efficient and sustainable infrastructure and its effective management in the long run for the next generation.

M. Yadav (✉)
Department of Advanced Aviation Management, Qatar Aeronautical College, Doha, Qatar
e-mail: myadav@qaa.edu.qa

T. Dhingra
School of Business, University of Petroleum & Energy Studies, Dehradun, India
e-mail: tdhingra@ddn.upes.ac.in

© The Author(s), under exclusive license to Springer Nature Singapore Pte Ltd. 2022
P. Jadhav and R. N. Choudhury (eds.), *Infrastructure Planning and Management in India*, Studies in Infrastructure and Control, https://doi.org/10.1007/978-981-16-8837-9_3

Keywords Airport infrastructure · Infrastructure · Airport terminals · Planning and management · Airport revenues · Technological interventions · Economic regulations · Tariff planning · Sustainable infrastructure

3.1 Introduction

The economic impact of air travel has mounted dramatically as a result of the remarkable expansion in air traffic. Its importance in the transportation of people, freight and the generation of employment cannot be overstated. Airport terminals have evolved into critical nodes in the technological and economic networks, as well as catalysts of localized socio-economic growth. A novel urbanization process is rapidly forming across these aerodromes as trade grows. The aviation industry helps to flourish by providing jobs, business, commerce and tourism possibilities. To accommodate rising needs, air transportation infrastructural facilities, notably terminal facilities and air traffic facilities, have been significantly extended and upgraded. There has been a temporal mismatch between the provision of infrastructural facilities and the need for them. In conjunction with ecological restrictions, financial constraints have slowed the growth of aviation infrastructures in recent years. However, it is widely acknowledged that terminals not only offer vital facilities for air carriers but also add value to the economic development of the vicinity and local community. The new concerns of terminal restructuring, safety and security must be addressed by aviation authorities. Acknowledging such tendencies, Government of India (GOI) opted to undertake significant systemic adjustments. The evolution of terminal infrastructures takes precedence, with a focus on providing effective and significant offerings at competitive rates. All Indian terminals were formerly under governmental control. As per global trends, the government is now dedicated to liberalizing terminal operations. Inviting corporate investors to use advanced technologies and managerial experience in this area is critical. Acknowledging the scope for development that the aviation sector provides, the corporate community is demonstrating a great enthusiasm for aviation projects.

The literature on this theme indicates that airports and airlines are entities, which share a common goal [1]. At the same time, they are in a relationship of service suppliers–customers [2]. Studies highlight that the factors which contribute towards the relationship between the airport and airlines are airport choice factors, airport's competition, airport infrastructure, airport operations, airport privatisation and airport valuation and vertical collusion [3]. The nature of airline become more competitive, resulting in shrinking yields and lower margins with elite product offering forcing LCC airlines to cut down the operating cost. This controlling pressure includes the station or airport charges which further pressure to the airport operators [4]. LCC airlines pushing the airport operators to lower down the airport charges and compensate the loss by the enhancing of non-aeronautical revenue [5].

The significant airport charges are regulated and monitored by the Airport Economic Regulatory Authority (AERA) in India. The authority adopts hybrid-till approach for all other major airports, including non-aeronautical contribution to 30–40% to offset the aeronautical charges. Literature reflects a concern regards to non-aeronautical revenue through single till dual till or hybrid till. Single till, i.e. cross-subsidizing the aeronautical cost entirely through non-aeronautical revenue approach resulting in reduced airport charges, is beneficial for airlines in reducing the operating cost wherever airports inclines towards dual till under which aeronautical and non-aeronautical cost and revenues are treated separately incentivizing the airport operators [6]. The hybrid-till model enlists the subsidizing of airport costs on basis of 30% of non-aeronautical revenue plus aero revenue as step taken by airport charges and tariff regulator Airports Economic Regulatory Authority across India [7]. "AERA favours the single-till approach as its more suitable for government airports and airline operators. There is strong debate over applicability of revenue approaches as Ministry of Civil Aviation MoCA is in favour of providing airport operator a better chance to flourish and develop providing powers to airport authority of India." [8]. Hans-Martin Niemeier pointed that passenger probability and the surplus is a crucial reason for boosting airport non-aeronautical revenues, not the airline itself [9]. David Starkie also pointed out that under dual till the charges would remain low as airports would garner higher revenue through upscaling in unregulated non-aeronautical revenue [1, 10]. Penned down that the airport operators are not able to cover the operating cost as few of the low-cost airlines tend to push the airport operator for the prices below the operator's marginal costs resulting limited options with airport operator to offer reduced landing charges and rely on economy of scope probability. Reference [11] mentioned that airports usually charge landing fees on the basis of aircraft weight, a fee per passenger that passes through the terminal as terminal user facility charges (TUFC), aircraft parking charges and charges for office space. Additional charges contribute to ground handling charges which airline can do for itself as well as provide ground handling services for other international airlines or by outsourced ground handling agency [12]. Barrett [11] highlighted that the low-cost airlines have a strong track record in delivering business even to less commercial viable airports offering non-aeronautical revenue sources such as catering and retail bundled services. Zou and Hensen highlighted the role of on-time performance (OTP) and punctuality, failing in maintaining the same, resulting in downscaling the market share and revenue of an airline. Fuel, crew and maintenance costs emerged as the high costs due to delay at airport excluding the buffer costs, i.e. airplane ownership, lease and rental, depreciation and insurance, etc. Zou and Hensen also define various costs as a result of schedule disruption, improper and off scale of operational efficiency at significant hubs in terms of missed flight connection, minimum connecting time (MCT) prolonged timeslot and cancellations of flights [13]. Zheng Lei, 2010 concluded that it is a rational decision for an airport operator to allocate capacity to the low-cost airline with cheaper average airport charges. Also, as low as 15% share in airport total traffic provides flexibility to the low-cost airline to negotiate for lower landing fees per passenger [14].Kindly note that reference citation [22] has been changed to [21] in the sentence 'Zou and Hensen, also defines various…'

The world is envious of India's aviation statistics. The sector has expanded continuously in double digits over the period from 2015, nearly doubling its extent, supported by high regional need, Ministry of Civil Aviation policies and corporate institution funding and involvement. According to the Directorate General of Civil Aviation (DGCA), an approximated 316 million travellers flew domestically (252 million) and internationally (64 million) in 2018–19, increasing 9.36% each year, making India one of the fastest-growing marketplaces [15]. The sector is poised to surpass more existing marketplaces if it keeps developing at this rate. According to the International Air Transport Association (IATA), the Indian airline sector will overtake nations, namely theUK, Japan, Spain and Germany, to emerge the third-largest airline marketplace in aspects of travellers by 2025, behind China and the USA [16].

Significantly, air transportation is evolving at a prompt pace. Affordable air fares have contributed to a spike in need in India's small cities, in conjunction with a strong gross domestic product (GDP) boom pushing higher air transportation expenditures. With greater domestic connection, air transportation is now becoming simpler, convenient and far more inexpensive for the common man, resulting in a significant shift in commercial openness. Domestic travellers load factors have reached new levels of 91% in February 2018, which is unsurprising. As long as fuel costs were reasonable, airline companies attempted to take advantage of the situation by introducing quitter and fuel-efficient modern airplanes. Subsequently, airline companies have been attempting to effectively optimize their expenses and capacity using hard and soft infrastructures, the scale of operations and new tech, resulting in lower operating costs and expenses.

In November 2019, IndiGo-one of India's highly lucrative air carriers commenced the incorporation of 50 modern turboprops into its portfolio. Indeed, the nation's latest Full-Service Carrier (FSC) Vistara Airline is looking to purchase approximately 50 narrow-body and 10 wide-body planes. One more significant participant in India's biggest domestic aviation marketplace SpiceJet made one of the biggest purchases for 205 airplanes from one of the largest commercial aircraft manufacturing company Boeing, valued at $22 billion [17]. Civil aviation is on the core list of India's constitution, which has a parliamentary framework. As a result, the domain of civil aviation falls under the parliamentary competence of the Government of India. The Aircraft Act governs commercial airlines in the nation, while the Aircraft Rules serve as the market's fundamental regulatory foundation. All air terminals were administered primarily by the Indian government until the International Airports Authority was established in 1972 and the Indian Airports Authority was established in 1986. To professionalize the administration of international air terminals, the International Airports Authority of India was established in 1972 and assigned control for international air terminals in metropolitan cities, namely Delhi, Mumbai, Chennai and Kolkata. The National Airports Authority of India was established in 1986 to oversee and administer the leftover public terminals, following the success of that authority. In 1995, the Airport Authority of India was formed by merging these two authorities. The major purpose of the merger was to expedite the construction, extension and modernization of all operative, terminal and freight infrastructures in accordance with global norms at all air terminals across the nation. Airport Authority of India currently has jurisdiction over 125 air terminals around the nation, comprising international, national and civil enclaves at certain defence facilities. The AAI is governed under the AAI Act, 1994. There have been several airfields around the nation that are controlled and administered by individual states or corporate enterprises, contrary to the air terminals under the AAI [18].

3.2 The Ownership Model Developments

Since 1995, the AAI has served as the nation's governing body for the construction and administration of airport infrastructures. The GOI planned in September 2003 to renovate and remodel Delhi's Indira Gandhi International Airport (IGIA) and Mumbai's Chhatrapati Shivaji International Airport (CSIA) using the public–private partnership (PPP) model. Subsequently, Airport Terminal 3, owned by DIAL, began operating in July 2010 which was constructed in only 37 months.

Under the UDAN plan, the number of non-AAI air terminals—controlled by private/corporate enterprises, local bodies and joint ventures—that has been opened for commercial flights has increased significantly over the years. This appears to have given the sector a more solid foundation. The administrative framework of operating airport terminals has changed dramatically during the last two decades.

3.3 Organization Structure in India

The growth of infrastructure begins with planning. The planning department has two main tasks: aerodrome planning and architecture. Airport planning corresponds to air-side planning and fiscal planning, whereas architecture planning is primarily focused on resource appraisal, novel resource progression and the design of buildings. In the technological examination of all established structures and the construction of novel structures, the two divisions of the planning department support one another.

3.3.1 The Key Responsibilities of the Architecture Planning Department Are as Follows

- Land identification for future terminals, land utilization and detailed design, capability evaluation, analysis of data and viability fact sheet production.
- Architectural planning of the aerodrome and aprons, recognition of regions to suit commerce needs, layouts of traveller and freight terminals.

- Cohesion with project managers and consultants throughout development, engineering analysis of client–agency ideas.
- Supervision of contractual airports projects through consulting.
- Modernization and expansion of city-side areas, as well as beautification/renovation of air-side amenities.
- Development of many other infrastructure complexes at terminals, comprising cargo containers, tech blocks and control buildings, institutional houses, Central Industrial Specific Force (CISF) and AAI staff houses, firehouses, health clinics, MT facilities and numerous auxiliary initiatives beneath Ethical Business Practices.

3.3.2 The Following Are the Primary Tasks of the Aerodrome Planning Department Are

- Aiding the DGCA and the Ministry of Civil Aviation (MoCA) in the areas of aviation system planning and airport planning, comprising land usage, runways, terminals, as well as other structures.
- Collaboration with aircraft companies, the DGCA, the National Economic Council (NEC) as well as other organizations. Describing the plans and specifications, preparing the progress reports, creating yearly strategies for airport improvements, a compilation of responses to inquiries from the Government, the Parliamentary Advisory Council, VIPs and others about terminal design and developments.
- Assessing the overall project's progress, giving tactical feedback to senior executives for tactical incorporation to achieve the company's goals, delivering plan feedback, such as plans and specifications, comprehensive project analysis, ecological approvals and so on, for project formation.
- Assessment of viability for airport terminal land acquisition, developing a revenue model for an airline terminal, assist the MoCA in formulating policies for airline development.
- Communicate with Air Navigation Services (ANS) divisions to integrate and synergize the Air Space Management Programme with the Airport Management Programme for peak efficiency development on the ground. Work with the Airline Security Division, Operations Team and Safety Department to improve airfield operating efficacy, security and safety.
- Supporting with the creation of the Airfield Strategic Design and land usage design, which includes city side modernization. Offer coherent aid for the formulation of terminal tariffs, as well as the maintenance and frequent revamping of statistics on airfields in the context of the airport directory.

- Creating a National Register of Major air terminals and updating it regularly, devising different ways for maximizing Air Side Capabilities, aiding in the deployment of Next Generation Systems in ANS by developing suitable necessary infrastructure and so on. Work with the local government to develop a city layout that is intertwined with the airport blueprint.

3.4 Airport Operational Aspect

Several civil aviation organizations use air terminals for their operations and infrastructure. These organizations serve a variety of purposes. Aside from direct consumers handling, air terminals must interact with a variety of exogenous stakeholders that, despite their distance, serve an essential role in an air terminal's accomplishment.

If you have ever been to an airport, whether as a traveller or a spectator, you have most certainly reached through these forms of ground conveyance, such as a vehicle, cab, bus or rail. As a departing traveller, the procedure might be repeated, with you coming by flight and departing via ground conveyance. One may get a description for an airport from the preceding remark: An airport is a venue where ground transportation and air transportation are exchanged. This concept presents intermodal operation; however, it does not address a scenario in which you come by plane and leave by plane without ever leaving the airport refer as transit passenger.

According to entrepreneurs, an airport terminal is an economic entity whose purpose is to offer travellers and traders amenities and utilities that satisfy their demands. Economic experts and policymakers see it as a vital commercial link with significant country's socio-economic benefits. These definitions propose the terms consumer demands and marketplace, as well as extrinsic advantages. Each of these definitions is correct and, to a certain level, applicable to air terminals, even though the primary role of an airport is to serve as a link between the ground and air modes of transportation involving several external agencies, private as well as government, playing pivot role such as central forces, immigration and custom, city police, ground handling agencies, cargo handling agencies, catering, food and beverages, retail and concessionaries.

3.5 Airport Classification

Landside, Terminal and Airside are the three primary regions that make up the airport. Let us take a closer look at each component of an airport.

3.5.1 Landside

Nearly every major aviation travel commences and textures on the ground. Moreover, not all journeys are taken by travellers individually; friends and relative may arrive at the terminal to drop or welcome travellers, workers may arrive for duty and cargo and commercial vehicles may be many. These vehicles need a well-designed roadway, ample parking and a secure link to the rest of the road and access system. Congestion in parking and roadways may create substantial inconvenience for travellers and terminal attendees. To address this qualm, technical developments such as the construction of specialized rail lines between the city and the air terminal, the usage of intra-airport autonomous vehicle transporters and computerized pay systems in parking areas and so on might be implemented. City-side amenities may be utilized to boost airport income by giving advertising spaces. Both the advertising business and the airport gain from these commercials since a firm must pay for utilizing the airport building for their marketing resulting in contributing non-aeronautical revenue for the airport operator. The airport operator is responsible for surface access operations, kerbside planning and design, evaluating the performance and service measures for landside along with capacity planning for vehicular traffic and pedestrian characteristics.

3.5.2 Terminal

The terminal complex is where travellers, baggage and goods are transferred from the ground mode to the air mode. The terminal is a distribution centre that has undergone continuous and significant technical improvements and update. Although initially terminals were constructed to keep travellers and staff safe from the weather, today's terminals are complex structures with nearly all comfort and emerging as commercial hub for airport operator. Most of the enhancements have been focused on safety, efficiency and convenience. Moving walkways are used at bigger terminals to minimize walking distances. To accelerate the operation and diminution overcrowding, computerized booking and self-check-in methods have grown ubiquitous enhancing capacity management with passenger satisfaction in terms of seamless and hassle-free travel experience.

Safety at air transport has since become a concern when civil aviation was first threatened. The terminal has previously provided a limited pre-clearance to the general admittance to the airplane: person screening was unparalleled, and luggage scanning was superficial. Terminals adopted procedures to prohibit unauthorized involvement in flight as security became more important. Traveller and luggage scanning, as well as terminal fences and limited airplane accessibility, were all done physically at first. Metal detectors, machine-readable passports and biometric authentication devices, among other things, are used in today's screening. Airport operator major operations includes terminal capacity planning, queue and flow management, identifying key performance indicators (KPI) to reduce processing time, trolley and floor management, stakeholder analysis and enhancing travelling experience with elite customer satisfaction.

3.5.3 Air Side

Aircraft were small, sluggish and flew under favourable weather circumstances in the initial periods of air transport. The only ground infrastructure needed was a tiny hangar, and they could fly off of grassed runways. Solid ground airstrips were to be built as airplanes became speedier and bigger, and the requirement for dependability grew. These airstrips were fitted with border illumination to enable aircraft to fly in poor visibility thanks to the modern navigational aids including Instrumental Landing System (ILS). Larger and broader airstrips/runways with better ground strength were necessary as the attributes of the airplane improved. However, powerful engines were developed after a few decades, and the airstrip size was gradually reduced and stabilized. Air Traffic Control (ATC) was established as the frequency of air traffic rose, particularly at and near air terminals: it was essential for a form of regulation to maintain travelling safer. ATC was initially created in the manner of Control Towers, tasked with preventing ground and air mishaps around terminals and ensuring continuous traffic movement. With improvements in technology, flight crews and ATC needed to communicate by radio and now the communication, navigation and surveillance (CNS) services has gone to another level. Airport operator major operations in terms of airside includes airside inspection including runway, follow-me operations, foreign object debris FOD identification, airside facility and capacity planning with strategic airside planning in terms of layout, runway usages, configuration and orientations. Modern airport operators are highly indulge in developing Airport Operation Control Centre (AOCC) as modern infra facilities to coordinate/streamline airport capacity and resource planning by airport resource allocations, updated information and communication, real-time data resource management and data-driven decision-making (D3M) process ultimately reducing the processing time and uplift the efficiency of not only airside but entire system.

3.6 Airport Operational Guidelines

The federal aviation authority [19] has introduced certain compulsory functions to be provided worldwide by all air terminals across the globe. These official procedures have been outlined in Section-D of Part 139, which deals with airport accreditation. The following are some of them:

1. Pavement Management—The uppermost section (surface) of an airstrip or carriageway is referred to as pavement. It is composed of either concrete or

asphalt. The pavement ought to have the appropriate weight-bearing capability, be suitable for flight operational activities and be capable of providing a smooth experience for airplanes. A pavement must meet the following minimal specifications: it must be free from external items (swamp, filth, etc.), have no fractures or openings within the surface and be flat and clean. For low light, the pavement should be properly lighted, and markers and signage must be precise. The air terminals are responsible for keeping the pavements in decent condition.

2. Safety Inspection—Inspections must be performed regularly to ensure the security of air transport. Inspections must be performed on places such as airplane parking (apron), airstrip, transfer points, towers, storage areas and fuelling stations, among other things. The purpose of the inspection is to ensure that the aforementioned places are free of obstructions, tyre waste, frost or snowfall, wildlife hazards and other hazards that might create significant disruption to the airport and the airplane.

3. Aircraft Rescue and Fire Fighting (ARFF)—It is obligatory for all terminals to offer this.

    ARFF's principal mission is to safeguard people. As ARFF distinguishes between the mortality and existence of all individuals onboard, its functionality at the terminal has to be at its maximum. ARFF has a three-minute timing restriction for responding to emergencies. Whenever incidents happen, they should be fully prepared and not sluggish.

4. Control of Ice and Snow—The accumulation of frost on the airstrip or sidewalk surfaces, or the airplane body, will influence the airplane's travel, i.e. the friction between the airplane's tyres and the pavement, causing issues with take-off and landings; snow on the airplane will create a lack of controlling the position and impair airplane efficiency. As a result, prompt clearance of ice and snow, proper choice of materials for pavement development, and notice to pilots if some pavement is inadequate are all necessary to minimize any hazard that might result in a mishap.

5. Bird/animal hazard management—Creatures fluttering in and near terminals have posed a significant hazard to planes. They inflict significant destruction to airplanes that could result in fatalities. Inside the air terminal's limits, the airport is accountable for managing the mobility and appearance of wildlife species. Installing buzzing sound devices to frighten creatures far from air terminals, informing ATC about any bird's movements and cutting down trees to decrease bird's population are some of the management measures that might alleviate this problem.

6. Additional functions, like NOTAM (Notice to Airmen), are available to provide updates on any dangers or changes to the system.

7. In order to convey wind flow directions to captains, traffic and wind direction signs must be put up. For commercial aviation activities at air terminals without a control tower, a real-time traffic indication and arrival runway signal must be installed across the airflow head on all airstrips with right-hand transportation systems.

3.7 Airport Planning Process

A master plan for an airport is a blueprint for its long-term growth. This blueprint is more than just a written representation of the airport evolution; it is also a stage-by-stage representation of the whole airport region's expansion, including both avionics and non-avionics requirements, as well as terrestrial usage nearby to the air terminal, with both monetary and technological ramifications. The elements of a standard airport system are depicted in the diagram below. The system's core elements are the ground and air sides. The main contact between the two elements is the airport building. The airport master plan's goals are shown below:

- Development of architectural installations at an airport and prospective property around the air terminal.
- Creating a schedule of objectives for implementing the plan's suggested stage growth and enhancement.
- Examine the project's technological, commercial and budgetary viability. Sustainability issues about airport functioning must also be addressed.
- Regulations and prospective aviation need for spending, growing obligations and managing land usage should be documented.

Table 3.1 summarizes the key aspects of the ICAO master plan guidelines [20]:

The specific airport master plan serves as the foundation for a long-term, detailed and collaborative development approach. The master plan emphasizes the airport intricacies and scale. The master plan frequently targets a specialized issue, namely rebuilding airstrips, evaluating obstructions, or upgrading navigational assistance or terminal arrival assistance. The master plan includes identifies structural upgrades, namely additional or longer airstrip, transfer points and apron extension.

The airport terminal planning method may be customized to fit the capacity of the terminal and the amount of budget allocated. The stages in the diagram can be merged to make the entire process easier (Fig. 3.1).

3.8 Guidelines by Airport Authority of India on Airport Planning

3.8.1 Introduction

The Greenfield Airports Policy [21] has indeed been published by the Govt. of India. Additionally, an Advisory Committee was set up to organize and oversee the different approvals necessary for the construction of Greenfield air terminals. These instructions are provided in accordance with the aforementioned policy and describe the method to be accompanied for the presentation, evaluation and acceptance of requests submitted for the establishment of an air terminal.

Table 3.1 ICAO master plan guidelines

Planning step	Description
Preplanning considerations	Coordination, planning, procedure, planning, organization, aims and policy objectives
Forecasting for development purposes	Needs, forecasting reliability, forecasting techniques and concepts, components and forecast demonstration
Fiscal measures and controls	Capital expenses include financial needs, funding sources and internal and overseas finance Revenue source, budgetary management and bookkeeping are all operating costs
Land assessment and selection	Site requirements, prospective site locations, variables influencing airfield design, initial design research, site assessment; functional, sociological and financial concerns, ecological research, viable site assessment, summary plans and income and expense estimates, summative examination
Airstrip and taxi strip	Measurements, sturdiness, airplane attributes, efficiency and airstrip size, as well as terminal layout and capacity
Aprons	Apron design, stall capacity, parking, maintenance and hanging aprons, lockers, safety, lodging in aprons
Air and ground navigation and traffic control services	Visual support, navigation systems and their structures, segmentation of important regions, air transportation facilities, facilities for inquiry and restart, interactions with apron controller
Passenger building	Fundamentals of planning, terminal traffic and functional quality, variables influencing the extent of services to be provided, infrastructure and need
Freight services	Location, type of construction and purpose, apron, infrastructure requirements, accessibility, parking, regulation and supervision are all factors to consider
Ground transport and internal airport vehicle circulation and parking	Automobile parking, personal and governmental transportation options, traffic statistics, interior carriageway layout
Airport operation and support facilities	Management and servicing, health clinic, interceptor service centres, power depots, water and waste management, weather services, airline staff housing, airplane repair, emergency firefighters and typical flying infrastructure are among the resources delivered. Fuelling stations for airplanes

(continued)

3 Emerging Scope of Airport Infrastructure: Case of India

Table 3.1 (continued)

Planning step	Description
Security	Airside safety measures: Streets, fences, separate parking spaces and a safe parking region
	Landside safety measures: traveller shelters and storage vaults facility

Source Self-compilation

Fig. 3.1 Airport planning process. *Source* Self-compilation

3.8.2 Institutional Structure

The Advisory Committee shall review all requests for the establishment of air terminals, regardless of type. The applicants will be needed to get permission from several administrative ministries/agencies; however, all ideas will be presented to the Advisory Committee. The Advisory Committee's administrator will be the Ministry of Civil Aviation, which would oversee ideas submitted by the Advisory Committee, holding Advisory Committee consultations and working with other decision-makers.

3.8.3 Applicability

All air terminals, helipads and airstrip upgrades shall be subject to these standards. These standards do not apply to air terminals constructed for personal usage, i.e. usage of the airstrip for non-commercial reasons.

3.8.4 Land Approval Granted by Advisory Committee

For land approval, the bidder intending to build a terminal would submit a proposal to the Advisory Committee in the manner attached at Annex-III. The submitted proposal will be transmitted for discussion to the AAI/DGCA. AAI/DGCA would conduct the land inspection and then send a recommendation to the Advisory Committee. The Ministry of Civil Aviation will give land approval depending on the assessment presented by the AAI/DGCA. If the planned land is developed in Controlled Airspace (Air Space beneath Ministry of Defence), the Ministry of Defence's permission will be sought as part of the land approval process. Within 3 months of receiving the proposal, the Advisory Committee would provide a decision on the Land Approval.

3.8.5 Assessment and "In Principle" Grant by the Advisory Committee

Following receipt of Site Approval, the appellant would submit a request to the Advisory Committee in the way outlined in Annex-IV. Requests for approvals from the Defence Ministry, Home Affairs and Finance ministry will be sent straightforwardly to all these agencies in a specified manner, with notification to the Advisory Committee's Secretary.

3.8.6 Guidelines

The relevant authorities will provide regulations for issuing permits. If necessary, these regulations will be updated periodically. The Ministry of Défense has released some recommendations for obtaining Ministry of Défense approval for the construction of Greenfield terminals. Annex-V includes the full recommendations. The applicants must file a request to the Ministry of Defence for a NOC in the manner prescribed in Annex-VI.

3.8.7 Permit by DGCA

After receiving the Advisory Committee's "in principle" clearance, the applicants must submit a request for an Airport Permit to the DGCA in a specified manner.

3.8.8 Roadmap for the Clearance and Assessment Processes

The GoI should make every effort to guarantee that all permits and approvals are granted in a reasonable timeframe. If a certain problem comes, the Advisory Committee will make every effort to stick to the recommended timeline:

(a) Land approval within 3 months of the time of the request.
(b) "In principle" permission: within 3 months of the time of the request.

Requests that must be forwarded to the Ministry of Civil Aviation or the Government will take longer than the period allotted above.

3.8.9 Monitoring

The Advisory Committee will keep track of projects that have been given "in principle" clearance. If no meaningful growth is achieved towards meeting defined goals within a specific timeframe, the Advisory Committee can decide to rescind the project's clearance.

3.9 Concluding Thoughts

Across the modern airport throughout the globe, the airport of the next generation has already been beginning to appear via innovative and immersive architecture. Comfortability and very well architecture techniques have indeed risen in popularity and are demonstrating to be fruitful, as evidenced by travellers happiness, comfort and satisfaction and air terminal's income. Air terminals should persist to accept chances to utilize effective approaches that emphasize the convenience of mobility, the overall user experiences and environmentalism as individual and aviation high-tech progresses.

Waiting time or processing time is the major concern of today's airport operators i.e. check-in time, security and screening time, frisking time, boarding time clubbed with turnaround time, taxing time and runway occupancy time (RoT) to build capabilities and generating revenues. The near-future air terminals will be all about movement. As screening turns "invisible" and facilities get innovative and customized, travellers will have a more personalized journey via air terminal, with progressively smooth movements and reduced borders among air terminals. Travelers may be allowed to utilize an NFC-enabled smartphone as a boarding pass to immediately unlock customs, airport lounges and departure doors, thanks to the secured NFC method. According to the current International Air Transport Association report, "self" bag tagging and "self" boarding have already been deployed in 115 locations throughout the globe.

Personalized smartphone technique also complicates the question of who governs travellers, investigating personalized branding and engagement with travellers in the quest for additional income opportunities both for airports and for the airline company. Seventy-eight percentage of airline companies intend to personalize information using straight distribution networks and data gathering traveller accounts in 2015. According to the latest polls, 61% of travellers are receptive to targeted mobile advertising provided they are appropriate and give them considerable choice. In addition, the traveller polls highlight the significance of travelling experiences and non-traditional architectural characteristics: the most desired airport amenity, according to 31% of travellers, is outside park space.

The perspective of the airport does not just alter, but also of the airport itself. Advanced technologies must evolve into smart buildings in order to satisfy societal and sustainable objectives. The airports of tomorrow will employ an "all of the above" sustainable approach, gathering its water and electricity resulting into energy savings, processing its garbage under waste management strategies and producing alternate bio-green fuel for aircraft. Several air terminals might advocate for a 24/7 model as they strive to improve capability under architectural growth restrictions; however, the possibility for greater capability must be balanced against overnight noise consequences and minimal need for late night or early morning flight hours. Finally, the futuristic airport will be a powerful mechanism that adapts its capabilities to millions of reliable data inputs on meteorological and travellers activities. A few experts believe that almost no cost control of privatized air terminals is necessary since the rewards for producing non-aviation income, such as subsidy and vehicle parking income, would prevent airport administration from imposing monopoly airfield's utility. Furthermore, [22] have demonstrated that, even after accounting for the additional benefit given by subsidy earnings, the airfield's utility prices of an uncontrolled revenue-amplifying airport are greater than that of a civil terminal operating beneath breakeven fiscal restriction. As a result, government agencies must carefully evaluate whether to apply any type of pricing restriction at privatized airports. The approach from government in terms of tariff fixation, regulations and policies regarding ownership and investment, airline airport sharing infrastructure, ease in airport infra financing and government grants with transparent legal framework will shape the future of smart and sustainable airport infrastructure for next generation.

References

1. Neufville, R. D. (2008). Low-cost airports for low-cost airlines: Flexible design to manage the risks. *Transportation Planning and Technology*, 35–68.
2. Barbot, C. (2013). Vertical collusion between airports and airlines: An empirical test for the European case. *Transportation Research Part E: Logistics and Transportation Review*, 3–15.
3. Yadav, M., & Dhingra, T. (2018). Recent developments in 'low cost carrier' research: A review. *International Journal of Business Excellence, 16*, 427–453. https://doi.org/10.1504/IJBEX.2018.096213
4. Graham, A. (2013). Understanding the low cost carrier and airport relationship: A critical analysis of the salient issues. *Tourism Management*, 66–76.
5. Zuidberg, J. (2014). Key drivers for differentiated airport passenger service charges. *Journal of Transport Economics and Policy*, 279–295.
6. AERA, A. E. (2019). *AERA annual report*. Government of India (GoI).
7. Chandra, J. (2021, August 22). *The hindu business*. Retrieved from www.thehindu.com: https://www.thehindu.com/business/Industry/explained-will-changes-in-aera-act-help-smaller-airports/article36038936.ece
8. MoCA, M. O. (2019). *Annual report*. Government of India GoI.
9. Peter Forsyth, C. G.-M. (2020). Covid−19, the collapse in passenger demand and airport charges. *Journal of Air Transport Management*.
10. Starkie, D. (2011). European Airports and airlines: Evolving relationships and the regulatory implications. *Journal of Air Transport Management*.
11. Barrett, S. D. (2004). How do the demands for airport service differ between full service carrier and low cost carrier? *Journal of Air Transport Management*, 33–39.
12. ICAO. (2009). *ICAO's policies on charges for airports and air navigation services*. ICAO.
13. Zou, B., & Hensen, M. (2013). Airport operational performance and its impact on airline cost. In *Modelling and Managing airport performance* (pp. 119–143).
14. Zheng Lei, A. P. (2010). Measuring the effect of low-cost carriers on regional airports' commercial revenue. *Research in Transportation Economics*, 37–43.
15. DGCA. (2019). *Annual report*. Government of India.
16. IATA. (2019). *Annual report*. IATA.
17. IBEF. (2020). *Indian aviation industry report*. Ministry of Commerce and Industry.
18. AAI. (2019). *Annual report*. Airport Authority of India AAI.
19. FAA. (2021). *Part 139 airport certification*. FAA.
20. ICAO. (2019). *National aviation planning framework*. ICAO.
21. AAI. (2008). *Guidelines for settingup of greenfeild airports*. AAI.
22. Tae Hoon Oum, A. Z. (2004). Alternative forms of economic regulation and their efficiency implications for airports. *Journal of Transport Economics and Policy*, 217–246.

Chapter 4
Rail Infrastructure—Journey Since Indian Independence and Beyond

Tarun Dhingra and Sanjeev Sharma

Abstract Indian Railways (IR) is the backbone of India's public transport network. One of the largest employers, it is the greenest, cheapest, fastest, and safest means of surface transport and caters to the needs of both the commuters and movement of goods for long distances in India. To achieve a 5 trillion dollar economy of India by 2024, the Task Force on National Infrastructure Pipeline (NIP) formed by the Government of India (GOI) has projected total investment of Rs. 111 lakh crore in crucial infrastructure sectors over five years FY 2020 to 2025. In railways, projects worth Rs. 13.68 lakh crore (12% of the projected investments) have been identified under this investment package. Such a significant investment calls for a review of the current status and milestones achieved by IR to plan further improvements for the sector. This paper attempts to review seven decades of IR to study various broad parameters like performance and safety, human resource, organization design and benchmarking and propose the framework for future rail infrastructure planning.

Keywords Indian Railways · Key performance indicators of Indian Railways · Rail transport in India and railway infrastructure in India

4.1 Introduction—Evolution of Indian Railways (IR)

Journey of IR started on 16 April 1853, when the first passenger train of the Great Indian Peninsula Railway left Boree Bunder station in Bombay (present-day Mumbai) for Tannah (present-day Thane). The investment in railways in India started with private ownership. Subsequently, there was a shift in ownership and operation starting in the 1870s, when the then Indian government began to construct and manage new railway lines. However, this period was short-lived, and in the 1880s, a hybrid government-owned but privately-operated structure emerged. This

T. Dhingra (✉)
University of Petroleum and Energy Studies, Dehradun, India
e-mail: tdhingra@upes.ac.in

S. Sharma
FA&CAO, Northern Railway, New Delhi, India

© The Author(s), under exclusive license to Springer Nature Singapore Pte Ltd. 2022
P. Jadhav and R. N. Choudhury (eds.), *Infrastructure Planning and Management in India*, Studies in Infrastructure and Control, https://doi.org/10.1007/978-981-16-8837-9_4

Table 4.1 Percentage share of major commodities carried by Indian Railways

Year	Coal	Iron ore	Cement	Food grain	Fertilizer	POL products
2015–16	65.45	70.78	36.99	18.13	87.53	16.54
2016–17	62.77	69.05	36.66	15.95	87.01	15.16
2017–18	62.83	66.68	37.43	15.26	85.46	14.87
2018–19	62.85	62.64	34.55	13.78	85.91	14.55
2018–19	60.03	62.72	32.88	13.10	84.23	14.58

Source Indian Railways Year Book 2019–20

public–private partnership model was the dominant organization in IR until the 1920s, when it gradually introduced complete nationalization. Around 1950, legislation was passed allowing the central government to take over many of the independent railway systems in operation and run IR as a government-owned departmental-run organization serving rail transport requirements across the length and breadth of India. IR transports every conceivable item, every component of the Indian economy like coal, steel, petroleum, fertilizers, and food grains. IR has been the principal mode of transport in the country. Table 4.1 shows the percentage of select significant commodities carried by the IR.

IR are the greenest, cheapest, fastest, and safest means of surface transport in India and cater to the needs of the commuters and movement of bulky goods for long distances in India. The comparative costs between railways, roadways, and waterways brought out by Niti Aayog in their recent study report are given in Table 4.2.

IR has the largest rail network in Asia having a multi-gauge and multi-traction system and the world's second-largest under one management. Some notable features of IR are as follows:

- Providing transport links through 67,956 route kilometres and 99,235 track kilometres is one of the largest railway systems in the world.
- A vast fleet with 12,729 locomotives, 57,060 coaches, 13,153 EMU/DMU/DHMU cars, 23 rail cars, 6372 other coaching vehicles, 293,077 wagons currently operates on an almost daily basis.
- Operates 22,648 trains and transports more than 1208 million tonnes freight and 8.09 billion passengers annually. Amongst 22,648 trains, people for commuting purposes and 8479 for freight use 13,169 trains.

Table 4.2 The Comparative cost of Railways, Roadways, and Waterways

Type of transport	Cost freight per tonne km (Rs.)
Roadways	2.58
Railways	1.41
Waterways	1.06

Source Niti Aayog report 'Goods on Move 2018'

4.2 Need for Study

IR is the backbone of India's public transport network catering to the needs of the commuters and movement of bulky goods for long distances in India. Over the years, the growth rate of revenues has been outstripped by the rate of increase in costs primarily due to the loss of market share in the profitable freight business, high cost of internally sourced products and services, increased fuel prices and staff costs and lack of flexibility in pricing. Investment in un-remunerative projects taken-up under socio-political consideration has further dented the finances of IR. As a result, IR's operating ratio (percentage of Operating expenses over Revenues) is hovering above 95%. The limited scope of internally generated resources and dwindling budgetary support from the Central Government and costly borrowings has led to low investment in infrastructure, resulting in capacity constraints, low speed, delays, and safety issues. The high-density network that connects metros has reached saturation, adversely affecting timeliness and quality of services. Poor infrastructure and services are resulting in loss of remunerative business for railways and further deterioration of finances. The way forward for railways to come out of this current scenario is investment in infrastructure and rolling stock to ensure availability of rake/seat on demand and better premium, high-speed goods, and passenger trains to sustain and augment its passenger and goods revenues.

Realizing the importance of Infrastructure development for growth of economy of the country, the Task Force on National Infrastructure Pipeline (hereinafter referred as NIP Report) formed by the Government of India (GOI) has projected total investment of Rs. 111 lakh crore in crucial infrastructure sectors over five years FY 2020 to 2025 to achieve a 5 trillion dollar economy of India by 2024. In railways, projects worth Rs. 13.68 lakh crore have been identified under this investment package.

Such a significant investment calls for a review of the present status of the IR and milestones achieved post-independence. This review is used as an input to suggest a framework for future rail infrastructure planning, which could be handy for creating a future performance baseline. For reviewing an organization as massive as IR this paper presents to study various broad parameters like performance and safety, human resource, organization design and benchmarking.

4.3 Methodology

4.3.1 Systematic Review of Literature

Systematic review of literature has been undertaken in this paper, with a thorough study of the IR. A systematic review requires collection and analysis of all articles pertaining to a particular research question keep the review unbiased and transparent. It is exhaustive and not as haphazard as traditional review [1–3]. The scope and the purpose of the systematic review are defined with the help of research question to seek out the relevant studies. Further appraisal of the studies is then carried out by applying the criteria of inclusion and exclusion. The relevant resultant studies are reviewed and analysed which leads to the synthesis of the results [1].

4.3.2 Question Formulation

The research questions have been formulated based on the researcher's intent to know the status of IR.

RQ: What Is the Present Status of Indian Railway?

The RQ requires us to review the studies on IR in totality including an in-depth study of the investment planning. All articles and other published documents pertaining to the research question were reviewed following the steps of systematic review outlined above.

Subject RQ: What Are the Review Indicators of Status of IR?

The review indicators have been arrived at with the help of relevant studies selected through the systematic review.

4.3.3 Article Search Procedure

Articles were searched from the electronic databases of Emerald, EBSCO, and Science Direct following the procedure of systematic review [2, 4, 5]. A detailed study was also undertaken with the help of policy documents (reports) pertaining to the IR. In addition, 'grey literature', viz. relevant reports, conference proceedings, etc., was also studied as its importance was emphasized by [1, 6]. The search was conducted under the selected databases with the terms—'Indian Railways', 'Key Performance Indicators of Indian Railways', Rail Transport in India, and 'Railway Infrastructure in India'.

Articles and reports were searched from 2005 to 2020 to make the review contemporary. The articles under which the search terms coincided with the keywords or the title were selected for detailed studies, else any paper even briefly mentioning about the IR would include the above-mentioned search term(s) and that would increase the total number of articles enormously, out of which only a very few would be actually focusing on the performance parameters. The initial search revealed more than 150 scholarly and peer-reviewed articles. These articles were filtered in three stages based upon the content. Although no document explicitly tells us the up-to-date status of IR, articles and reports mentioning the performance have been retained. Finally, after all the inclusions and exclusions, eight published articles besides Indian Railway Annual Reports, NITI Aayog's Reports on IR were found to be of substantial value for the intended research to ascertain the present status of IR and its investment planning and its impact on the rail infrastructure in India. Each selected article and reports were thoroughly read and analysed by the authors for its relevance to the topic.

4.3.4 Literature Review

Given the need for sustenance and growth of the national transporter for the Indian economy and national integration, several studies have been made. The brief highlights of a few relevant exploratory studies carried out during the last 10 years are brought out in Table 4.3.

4.4 Research Design

Objective: The present paper aims to study the key performance indicators of IR, analyse the strategies adopted, and suggested way forward for expanding the rail infrastructure and improve its financial health, with the help of information available in the public domain.

Data: IR publishes various reports and statements, including the Indian Railways Year Book, which contains information on the railway network (for each zone in addition to the entire system), railway infrastructure and expenditure thereon, its assets (e.g. tracks, electrification, rolling stock, traction, passenger business, and freight operations), and asset utilization. The Indian Railways Year Book and the Annual Statistical Statements from different years, the Reports of Niti Aayog, Economic Survey, and the findings from the exploratory studies done in last fifteen years on IR were our main sources of reference data and information.

Based on the above, the study has been divided into the following sections:

I. Organizational Structure in Indian Railways.
II. Physical performance including safety and security since 1950–51 to 2019–20 including comparison with a few Foreign Railways.
III. Financial performance since 1950–51 to 2019–20.
IV. Social service costs of IR.
V. Human resource, planning and value addition by the human resources in Indian Railways since 1950–51 to 2019–20.
VI. Analysis and findings—
 - Current challenges
 - Way forward
 - Investment planning
 - Financing and current status
 - Conclusion.

Table 4.3 Summary of exploratory studies on IR

S. No.	Author	Title	Year	Objective	Inference	Theme
1	[7]	Optimizing Indian Railways infrastructure by AI	2021	To study potential contributions of AI in improving the operation and profitability of Indian Railway System	To meet the higher demand, IR has prioritized capacity building by increasing the number of coaches per train, running more trains, and building more tracks involving high infrastructure costs. The use of AI in IR operations will improve the efficiency in train operation	Artificial intelligence; Indian Railway; signalling, maintenance
2	[8]	Expense-based performance analysis and resource rationalization: case of Indian Railways	2020	To evaluate efficiency performance of different railway zones and to suggest cost-saving strategies	Indian Railways has been striving hard to bring down the costs and improve upon the operating ratio. The results highlight the inefficiencies of different zones and data envelopment analysis (DEA)-Malmquist index show that technological innovation and staff management are essential factors in cost reduction	Operating and working expenses, input/output parameters
3	[9]	Digital social media: enabling performance quality of Indian Railways services	2018	Study of causes of customer dissatisfaction and technology intervention and managerial challenges to meet operating challenges and the way forward	The scale of mammoth operations by IR translates to humungous everyday challenges. Customer dissatisfaction is prevalent despite subsidized travel fares and multiple initiatives. The technology advancements can help managing large-scale public transport operations	The scale of operations, infrastructure, operational challenges, customer dissatisfaction, technical advancement

(continued)

Table 4.3 (continued)

S. No.	Author	Title	Year	Objective	Inference	Theme
4	[10]	Financial turnaround of Indian Railways (A)—Emerald	2012	The challenge faced by IR with its peculiar transport mix and limited resources; reason for the downtrend of IR finances; and innovative strategies adopted to get back to the path of growth-leading to the historical turnaround of IR	The main reasons for the poor performance of IR were severe competition from other modes of transport, rigid pricing, investment in un-remunerative projects, and other such practices. Various management experts recommended restructuring/corporatizing, reorganization, increasing passenger fares, unbundling non-core activities, downsizing, and outsourcing. However, IR met the challenges and attained unprecedented growth in traffic and earnings through certain strategic decisions	IR transport mix, operational constraints, challenges, strategies for turn around

(continued)

Table 4.3 (continued)

S. No.	Author	Title	Year	Objective	Inference	Theme
5	[11]	Financial turnaround of Indian Railways (B)—Emerald	2012	The impact of the global slowdown on the economy and turnaround phase of IR and discuss the strategies to return IR to the growth path	Indian Railways generate resources for its development expenditure and fully cover its operational costs in sharp contrast to most world railways that depend on a subsidy for operations and development expenditure. The constraints of fixed expenditure, essentially comprising staff-related expenses and fuel costs, make it challenging to achieve the target. The study analysed the influence of global developments on the Indian economy after 2008–2009, implementing the sixth pay commission's recommendations on IR's staff costs and future strategies and outlook	Global slowdown, impact on the economy, transport sector, investment plan and recovery strategies
6	[12]	Dedicated freight corridors (DFC) will boost rail productivity	2006	Need and benefit of DFC	DFC to be a game-changer to facilitate development of high-performance freight railways system capable of higher axle-loads, segregation of freight, and passenger services, allowing existing mixed-traffic routes to be optimized as high-quality passenger corridors	Existing infrastructure, bottlenecks, inefficiencies and costs

(continued)

Table 4.3 (continued)

S. No.	Author	Title	Year	Objective	Inference	Theme
7		Cost-based indexing of fare and freight to fuel pricing; innovative pricing policy for IR—cost management	2016	To develop a methodology that offers possible changes to the fare and freight structure to offset proportional changes in fuel prices	IR can adapt to increase revenues within the constraints of social responsibilities by enacting a suitable tariff linking with fuel price to cover both the cost of operation and the replacement of assets and business growth. To combat its current financial challenges, IR should focus on increasing yield/PKM in passenger services and acquiring additional traffic streams in goods services. Any future increases in rates should be cost-based in view of stiff competition from road. Besides, IR should extend electrification of the routes and use energy-efficient technology	Cost-based indexing of fare and freight to fuel pricing, tariff policy, financial challenges, fuel adjustment component
8	[13]	Profitability and leverage analysis of Indian Railways-impact of cost-based indexation—cost management	2017	To analyse impact of cost-based indexation on profitability and leverage of IR and cost effectiveness of electrical traction	The application of fuel adjustment component (FAC) in fare and freight would improve IR's profitability and leverages. Further, the electrical traction is cheaper and environmentally friendly. IR needs massive investment in electrification	BEP, operating and financial leverages, fuel adjustment component

4.5 Organizational Structure of IR

IR is governed by a three-tier vertical organization structure comprising Ministry, Zonal Railway, and divisional setups for effectiveness and efficacy in management of its operations. Each zonal railway is made up of a certain number of divisions covering stipulated area of operation and a divisional headquarter. There were nine zones, and this structure had not changed much for four decades. In 2002–03, 7 new zones had been created to achieve better administrative control, giving 16 zones. Besides zones, IR also has its production units producing rolling stocks, viz. locomotives, coaches, wagons etc., apart from the components.

The delegation of authority flows from top to bottom and the Ministry of Railways under the Government of India controls Indian Railways. The Union Minister, who is generally supported by one/two Ministers of State (Railway), heads the Ministry. The Railway Board consisting of three Director General (HR, Railway Protection Force and Railway Health Service), four Members (earlier eight), and a Chairman and Chief Executive Officer, reports to this top hierarchy. The railway zones headed by the General Managers report to the Railway Board. Apart from the Zonal Railways, IR also has its own Metro network at Kolkata, Mumbai, Chennai, Delhi, and production units manufacturing rolling stock and components to meet the requirement of the Zonal Railways. There are 68 divisions in IR. In addition IR also has its own research and development wing in Research, Designs, and Standard Organization (RDSO) functioning as the technical advisor and consultant to the Ministry, Zonal Railways, and Production Units. Besides, the Ministry of Railways also has sixteen undertakings under its aegis.

4.6 Indian Railway's Performance-Physical

4.6.1 Railway Network

The route kilometres in IR have increased to 67,956 during the year 2019–20 from 53,596 route kilometres during the year 1950–51, showing a growth of 127% during the above study period (Exhibit 4.1). Rate of expansion of route kilometres in the first 60 years beginning 1951–2011 averaged around 176 kilometres per year (10,577 km). During 1990–91 to 2000–01, the average annual expansion was a mere 66 km. In 2017–18, it plunged to a mere 17 km. The total growth of route kilometres in last 69 years is 14,360 km reflecting an average annual increase of around 208 km. This rate of growth is too low a rate for a railway line penetration in a country of India's size, more so because large parts of the country, especially the hilly regions still lack rail transport facilities. There is an unending demand for increasing the network. There is a throwforward of investments in sanctioned projects amounting to Rs. 10.80 lakh crore to end of 2019–20 (into 2020–21) (Exhibit 4.2). Expansion of the railway infrastructure is urgently needed to remain competitive vis-à-vis other transportation modes. Investments in the rail infrastructure have remained subdued as compared to power and road sectors. As per the NIP Report, between the financial year 2012–13 and 2016–17, the share of the railway's sector investment in the overall infrastructure investment was approximately 8%, rising at a CAGR of about 23%, though it picked up pace since 2015–16 as may be seen in Table 4.4.

During FY 1950–51 to 2019–20, freight loading increased from 73.2 million tonne to 1208.4 million tonne (1651% rise) and passenger kilometres from 66.5 billion kilometres to 1050.7 billion kilometres (1580% increase), but route kilometres grew only 127% (Exhibit 4.1). Of 247 sections, 161 (or 65%) are running at 100% or above line capacity on high-density network (HDN) routes, viz. Delhi–Howrah, Delhi–Mumbai, Mumbai–Howrah, Howrah–Chennai, Mumbai–Chennai, Delhi–Guwahati, and Delhi–Chennai (Source: PIB). Increased congestion and saturated usage of line capacity reduce the interval between two consecutive trains runs on the same route, thus increasing the chances of collisions on bustling stretches. It also leads to a reduction in the average speed and service delivery of passenger and freight trains. Besides, over-utilization of the existing network also harms safety. Most casualties happen due to derailments caused by defects in the track or rolling stock. There is a throwforward of Rs. 84,146 crore for track renewal. As per NIP report, out of 68,000 km of track, over 7000 km is more than 30 years old and needs immediate replacement. Early installation of anti-collision technology and advanced signalling systems to improve passenger safety is also required. Demand for new lines too has increased exponentially, with the supply-side struggling to keep pace. Over-utilization of many routes has also led to gradual wear and tear of tracks and decline in the average speed of trains, necessitating significant investments in network expansion, track renewal and decongestion projects rapidly.

4.6.2 Traction

The IR that started its journey with coal as its fuel is now using HSD and electricity for operation of the trains. Steam traction is confined to the running of heritage/tourist trains and a few narrow gauge hill sections. The electrification network in IR has increased to 39,329 km during the year 2019–20 from 388 route kilometres during 1950–51, showing a 101.4 fold growth during the above study period. The electrified route comprises 57.87% of the total network (Exhibit 4.1). The double line Golden Quadrilateral linking the four metros, viz. Delhi, Mumbai, Kolkata, and Chennai, has been fully electrified. 67.3% of freight traffic and 59.6% of passenger traffic is hauled on this electric traction.

Electric traction cost is 38.4% of the total traction fuel cost on IR. Therefore, IR has planned to electrify balance BG rail routes by 2021–22 to achieve 100% electrification of all its broad gauge rail routes.

Table 4.4 Investment in rail infrastructure (Rs. in crore)

Source	2011–12	2012–13	2013–14	2014–15	2015–16	2016–17	2017–18	2018–19	2019–20
General budgetary support	21,073	25,234	28,174	31,617	37,516	45,232	43,418	52,838	67,477
Internal resources	4198	10,007	10,590	16,057	16,938	12,125	3070	4663	1685
Extra budgetary support	14,790	15,142	15,225	11,044	39,066	52,578	55,498	75,876	78,901
Total	45,061	50,383	53,989	58,718	93,520	109,935	101,986	133,377	148,064

Source Indian Railways Year Books 2011–12 to 2019–20, budget documents 2011–12 to 2019–20

Rolling Stock: Locomotives

From loco holding of 8209 comprising 8120 steam, 17 diesel, and 72 electric locomotives in 1950–51, IR now has 12,147 locomotives including 39 steam, 6049 diesel, and 6059 electric locomotives. The tractive effort of these locos has gone up to 39,037 kg in the case of BG and 16,454 kg for MG as in 2019–20 from 12,801 kg in BG and 7497 kg for MG in 1950–51 (Exhibit 4.3).

Rolling Stock: Wagons

IR has a fleet of 293,077 wagons comprising 67,011 covered, 169,871 open high-sided, 17,473 open low-sided, 23,664 other types, and 15,058 brake vans/departmental wagons (Exhibit 4.4). These different types of wagons, e.g. such as covered, open, flats, well wagons, four-wheeler tank wagons, cater to transport requirements of various kinds of goods. While there has been a considerable decrease in the size of the total wagon fleet on IR since the eighties, wagon capacity has improved owing to technological improvements in wagon design and rerating of carrying capacity of wagons.

Rolling Stock: Coaches

The number of passenger coaches has increased from 19,628 in 1950–51 to 76,608 in 2019–20 to meet the ever-increasing passenger travel demand. The design of passenger coaches has undergone a qualitative change in physical structure, capacity, safety, comfort, eco-friendly, and continuous process (Exhibit 4.5).

Stations

The public is connected to IR with stations. The number of railway stations has increased to 7325 during the year 2019–20 from 5976 during 1950–51 during the study period (Exhibit 4.1). Opening of new railway stations is required to cater to the commuters' needs and tap the prospective demand in the various places in the future.

Goods Sheds

The silver spoon for IR is its goods traffic segment. It has about 1150 goods sheds, and to increase the coverage, it has allowed a no. of private and assisting sidings (1400 private sidings). Private freight terminals have also been allowed to achieve rapid development of freight terminals with private investment. At present, there are about 59 private freight terminals. The objective is to provide efficient and cost-effective logistics services with a warehousing solution to end-users.

Productivity

IR is in the business of selling transportation and its productivity is measured in terms of the traffic output, i.e. Passenger Kilometres (PKM) in case of passenger traffic and Net Tonne Kilometres (NTKMs) in case of the goods traffic. The total passenger output of IR has increased to 1157.1 billion PKMs during 2018–19 from 66.51 billion PKMs during 1950–51, showing a 17.4 fold growth during the above period (Exhibit

Table 4.5 Relative position of passenger transpoted by Major countries in 2017

Millions PKMs	USA	China	Russia Federation	India	Germany
2016–17	10,660	685,213	122,920	1,149,835	77,500

Source NIP report, Word Development indicators, and IR Yearbook 2017–18

Table 4.6 Relative position of goods transpoted by Major countries in 2017

Million tonne km	USA	China	Russia Federation	India	Germany
2016–17	2,445,132	2,146,466	2,491,876	620,175	70,614

Source NIP report, Word Development indicators, and IR Yearbook 2017–18

4.1). It dropped to 1050.73 billion PKM during 2019–20 due to the adverse impact of COVID. Passengers carried which were 1284 million in 1950–51 have increased to 8439 million in 2018–19 showing 6.5 fold growth during above period. The number of passenger carried also dropped in 2019–20 to 8086 million due to the impact of COVID on travelling. Relative position of passenger transported in 2017 (million passenger km in USA, China, Russia, India, and Germany is indicated in Table 4.5).

In case of freight traffic, the productivity of IR has increased to 738.52 billion NTKMs during the year 2018–19 as compared to 37.56 billion NTKMS in the year 1950–51 showing 19.7 fold growth during the above period. The performance had a setback in 2019–20 with 707.66 billion NTKM (Exhibit 4.1). The higher productivity reveals the higher output and is a significant efficiency parameter. Freight loading had increased 16.7 fold from 73.2 million tonnes in 1950–51 to 1221.48 million tonnes in 2018–19. Freight loading dropped to 1208 million tonnes in 2019–20 due to the adverse impact of COVID. Relative position of goods transported in 2017 (million tonne km in USA, China, Russia, India, and Germany is indicated in Table 4.6).

IR has envisaged Vision 2024 to achieve targets of 2024 million tonne freight loading by 2024. For this, super critical and critical projects have been identified for resource allocation for throughput enhancement faster. Supercritical projects are targeted for completion by December 2021 and critical projects by March 2024 (Exhibit 4.11).

4.6.3 Safety Performance

Safety and security are essential to gain the continued patronage of existing commuters/users and tap the potential demand. IR takes care of the safety and security of commuters through its own police force and the state police services. The number of accidents in IR has reduced to 54 during the year 2019–20 from 891 during the year 1970–71. The number of train accidents has also declined by about 94% during the above study period indicating a healthy trend (Exhibit 4.6). The safety index in IR reflected through train accidents per million km is 0.05 during the year 2019–20 as compared to 1.91 during 1970–71 and shows substantial improvement during the above study period.

4.6.4 Indian Railway's Performance-Financial Performance

The source of revenue of IR comprises revenue from passenger services, other coaching (parcel, etc.), goods services, and sundry other (non-fare box). The position of revenues since 2012–13 is given in Table 4.7.

Gross revenue of IR has increased to Rs. 190,507 crore during the year 2018–19 from Rs. 263.30 crore during the year 1950–51, showing a 723.5 fold increase during the above study period (Exhibit 4.7). However, the gross revenue has dropped to Rs. 174,694.69 crore during 2019–20 due to COVID-related slowdown and lockdown. The total working expenses, including depreciation and pension provisions, has increased 865.5 fold to Rs. 186,734 crore in 2018–19 from Rs. 215.74 crore in 1950–51. The total working expenses in 2019–20 reduced to Rs. 173,105.07 crore due to less appropriation to pension fund (Rs. 20,708 crore provided against Rs. 48,350 crore required) a result of COVID-related resource gap. Had the amount needed of appropriation were made, the total working expenses would have increased to Rs. 202,336.69 crore, showing 937.9 fold increase since 1950–51.

The net revenue, which reflects the contribution of IR to the government exchequer, has increased to Rs. 3774 crore in 2018–19 from Rs. 47.56 crore during the year 1950–51, showing a 79.3 fold growth during the above period. The net revenue reduced to Rs. 1589.62 crore in 2019–20. However, had the required amount of appropriation to the pension fund were made, there would have been a deficit of Rs. 26,052.38 cr. It is pertinent to mention that net revenue increased to Rs. 19,228.48 in 2015–16 (404.3 fold increase since 1950–51) but came down to the present level due to the impact of implementing recommendations of 7th Central Pay Commission. Further, during 2019–20, it has been impacted adversely by COVID. The higher Net Revenue depicts the greater financial soundness. The share of rail GDP in overall GDP of the nation has remained close to 1% in the last five years.

The total investment in IR has increased to Rs. 640,408.27 crore in 2019–20 from Rs. 855.2 crore during the year 1950–51, showing a 748.8 fold growth during the above study period. The Return on Investment (ROI) in IR was 1.08% during 2018–19. It dropped down to 0.29% in 2019–20 as a result of COVID-related resource gap.

The operating ratio (% of total working expenses over total revenue earnings), which depicts the efficacy of the operational management of IR, was at 97.3% during the year 2018–19 as compared to 81% during 1950–51. It increased to 98.36% in 2019–20. However, had required amount of appropriation to the Pension Fund were made, the operating ratio would have been 114.19%. A higher operating ratio curtails IR's ability to plan fresh investments through internal accruals. The operating ratio was 90.5%, and return on investment was 7.0% during 2015–16. The increase in operating ratio and decrease in return on investment were mainly due to an increase

Table 4.7 The position of Revenues from 2012-13 to 2019-20

		2012–13	2013–14	2014–15	2015–16	2016–17	2017–18	2018–19	2019–20
1	Gross traffic receipts (2 + 3 + 4)	123,733	139,558	156,711	164,334	165,292	178,725	189,907	174,357
2	Freight revenue	85,263	93,906	105,791	109,208	104,339	117,055	127,433	113,488
3	Passenger revenue	31,323	36,632	42,190	44,283	46,280	48,643	51,607	50,669
4	Other traffic revenue	7147	9020	8730	10,843	14,673	13,027	11,407	10,200
5	Miscellaneous receipts	2448	3656	4307	4046	90	204	601	338
6	Internal revenue (1 + 5)	126,180	143,214	161,017	168,380	165,382	178,929	190,507	174,695

in pay and pension on implementation recommendations of successive Central Pay Commissions on the one hand non-commensurate enhancement of fare and freight on the other. The COVID-related impact has dampened the revenue during 2019–20. Higher operating ratio curtails IR's ability to plan fresh investment through internal accruals.

4.6.5 Social Service Costs of IR

There is a long tradition of IR to provide essential public service to those unable to pay fully as part of its social service obligations. These social service obligations include loss in low-rated commodities of use by common people, concession in passenger fare, losses in EMU suburban service, losses in uneconomic branch lines, losses on strategic lines, losses on the parcel, luggage, postal, and catering services, losses on account of pricing of fares below cost, etc. The cost of these obligations during 2019–20 was Rs. 45,542 crore (Exhibit 4.8). The freight users and passengers travelling in higher classes generally cover the loss of fare from the poorer sections.

4.6.6 Human Resources in Indian Railway

Equipment, procedures–those things can be duplicated. Human capital is the only area where companies can differentiate themselves.—Meldron Young.

The total staff strength of IR is 12.54 lakhs in 2019–20 as against 9.14 lakhs during the year 1950–51, showing a growth of 137% during the above study period. The human resources on IR are classified in four broad groups/categories, viz. Group A&B, consisting of Officers Cadre, group C comprising of supervisors and subordinate staff, and group D as the support staff. Group D now stands subsumed in Group C.

The value addition made by the human resources arrived as net of gross revenue per employee and cost per employee has increased to Rs. 453,993 during the year 2018–19 from Rs. 1636 during 1950–51 showing a 278 fold growth during the above study period (Exhibit 4.9). Value additional decreased to Rs. 162,459 in 2019–20 due to COVID-related revenue loss.

The human resource cost on IR has increased to Rs. 154,214.71 crore during the year 2019–20 from Rs. 113.80 crores during the year 1950–51, showing a 1355.1 fold increase during the above period. The human resource cost amounts to 73% of the total cost on IR (Exhibit 4.9).

IR is following a strategy of trimming the human resource strength and has pruned down its staff strength from 16.52 lakhs (1990–91) to 12.54 lakhs (2019–20) showing a downsizing by 3.98 lakhs (24%) during the above study period. It translates to cost reduction to the tune of Rs. 48,979 crores (3.98 lakhs × Rs. 1,230,641, i.e. average human resource Cost) during 2019–20. It is one of the major avenues to generate more net revenue in IR. Thus, IR should focus on the core activities and hive off the non-core activities to private sector/specialized business process outsourcing (BPO) and knowledge process outsourcing.

4.7 Analysis and Findings

4.7.1 Current Challenges

Given the constraints in which IR is operating, its performance is satisfactory and its contribution to national development and integration during the seven decades is praiseworthy. However, business as usual is no more a saner choice for IR. The operating ratio of above 95% manifests that its capability to generate operational surplus is low. Ever-increasing expenditure on staff and pension is not controllable being a government entity. As a result, its capacity growth is increasingly being funded through costly borrowings worsening the financial situation of railways further. Improving operational surplus through fare and freight hikes also has limited scope for railways because:

- Second-class passenger traffic, which contributes to 50% of the total passenger revenue in the passenger segment, is cheaper than both road and air travel. The increasing fare in these services has socio-political consideration as it serves a relatively poorer section of society. The loss is presently being met by cross-subsidization to upper-class passengers and freight traffic. The passenger business made losses of more than Rs. 45,542 crore in 2019–20, classified as social service obligations (Exhibit 4.8). The subsidy ought to be ideally borne by the government/state government (for suburban traffic) is being borne by Railways, which is eating its revenue earning capacity substantially.
- The share of railways in total passenger traffic has declined from 74% in 1950–51 to 6.9% in 2015–16 (Exhibit 4.10). The upper-class passenger traffic is facing stiff competition from low-cost budget airlines and AC buses. Thus, an increase in upper-class fares is fraught with losing this class of traffic to other modes of travel, both faster and convenient.
- IR's freight rates are already higher than other modes of transport for several commodities. Consequently, the railway is losing its traffic to the road sector, which is relatively cheaper and logistically more convenient due to better last-mile connectivity. The share of railways in total freight traffic has declined from 86% in 1950–51 to 24% in 2015–16 (Exhibit 4.10). At present, its freight basket is limited to certain bulk commodities, and heavy dependence on coal

transport poses a risk to the business due to industry preference to pit head-based powerhouses. The freight traffic should be focused on tapering new traffic streams and recapturing the traffic lost to road.
- IR is facing stiff competition from roads in freight transportation and airlines in passenger transportation. While growth in passenger rail traffic is steady, airline and road traffic have seen substantial growth in recent years. To remain competitive vis-à-vis other transportation modes and provide optimum service to passengers/freight, IR needs to upgrade its current infrastructure. Greater private sector engagement in this sector by creating a conducive policy and regulatory environment is the need of the hour.
- Lack of rake availability for cargo movement is also a major issue for freight transportation by rail. It is further aggravated in the case of unregulated sectors such as steel and cement. In these sectors, rail transport is a significant component of the supply chain for sourcing raw materials and placing finished products in the market. In the existing scenario of lack of rake availability, end-users switch to other modes of transport, leading to a sub-optimal modal share of railways.

4.7.2 Way Forward for IR

Way forward for railways to come out of this current scenario would be to invest in dedicated freight corridors and logistics that could bring back its lost freight traffic, availability of rake/seat on demand and better premium, high-speed goods and passenger trains to sustain and augment its passenger and goods revenues. The number of passenger vehicles should further increase with more comforts, safety, and security. The superior services could be priced higher than average freight and fare. Further, the challenge would be to make these services cost-effective for the user.

4.7.3 Investment Planning

To achieve 5 trillion dollar economy of India by 2020, Task Force on National Infrastructure Pipeline (NIP) formed by GOI under Chairmanship of Secretary Expenditure, on general and sector reforms by the Centre and states has projected total investment of Rs. 111 lakh crore in key infrastructure sectors over 5 years FY 2020 to 2025. The emphasis would be on ease of living-safe drinking water, access to clean and affordable energy, health care for all, modern railway stations, airports, bus terminals, and world-class educational institutes. Sectors such as energy (24%), roads (18%), urban (17%), and railways (12%) amount to around 71% of the projected investments. Other sectors included are irrigation, mobility, education, health, water, and the digital. Vision 2025 for the IR is summarized below:

Status	Vision 2025
• The low modal share of Indian Railways in freight traffic at 33%22 • Two DFCs of total length ~3360 km under implementation • High-speed railway network—NIL • Two stations being developed on PPP basis—Gandhinagar and Habibganj, others under development	• The increased modal share of Indian Railways in freight traffic at >40% • EDFC and WDFC23 fully operational with the commencement of construction of other planned DFCs such as East–West DFC, North–South DFC, East-Coast DFC, and Southern DFC • Mumbai–Ahmedabad HSR to be functional and other identified HSR projects at the implementation stage • Higher private participation: 30% of net cargo volumes and 500 passenger trains privatized; 30% of 750 stations privatized; rolling stock from the private sector
• 46% of the existing railway network has been electrified as of 31 March 2018 • High network congestion: 50% of the 227 sections on the high-density routes are operating at a capacity utilization of above 120%	• 100% of the existing railway network electrified • Optimum utilization of existing rail network—fewer train delays, due to doubling/tripling/quadrupling of sections on high-density corridors—completing multiple tracking works of 14,100 km on high-density network and highly utilized network
• Limited focus on safety and security aspects • Customer experience needs to be improved given the lack of basic amenities and frequent delays • Average accidents per year: ~113 for 2015–17	• Focus on safer travel: railway accidents to reduce drastically • Improved customer experience with high-quality amenities with modern stations and high-quality catering • Smart coaches with on-board infotainment, Wi-Fi, CCTV, fire and smoke detection facilities, tea and coffee vending machines • The increasing average speed of freight trains to 40 km/h by 2024—due to investment in better wagons, ROB, RUB, flyover//bypasses, track upgrades, up-gradation to high-power locomotives

Source Indian Railways Annual Reports and Accounts, Bureau of Transportation and Statistics

Under NIP, a total capital expenditure of Rs. 1,367,563 crore would be made between FY2020-25 in IR by Centre and States in about 724 identified projects. Out of the 724 projects, 697 projects worth Rs. 11.97 lakh crore will be implemented through EPC mode (Engineering, Procurement, and Construction). Twenty-seven projects worth Rs. 1.61 lakh crore will be implemented through public–private partnership (PPP) model. The project includes DFC, Mumbai-Ahmedabad High-Speed Railway, Semi High-Speed Railway (Delhi-Mumbai and Delhi-Howrah), Signalling, Additional lines, New Lines, Gauge Conversion, Loco, and Loco Sheds, Rolling Stock, Energy Management, Railway Electrification, Track Machines, Telecom, Terminal Facility, ROB/RUBs, Bridges, Suburban, Track Renewal, and Private Train Project. Priority has been given to projects with high traffic potential. A summary of these identified projects is highlighted in Table 4.8.

Table 4.8 Summary of priority project with high traffic potential

Category	No. of projects	Capex over FY20–25 (Rs. crore)
New lines/gauge conversion	259	440,072
Capacity augmentation	266	247,985
Dedicated freight corridor	7	166,171
Rolling stock	31	275,539
High-speed rail	2	110,647
Others (coach and freight terminals and maintenance sheds)	159	118,406
Total	724	1,358,820

Table 4.9 Capital expenditure over FY20 to FY25

Rs. in crore	FY20	FY21	FY22	FY23	FY24	FY25	Total
Centre	132,463	260,811	307,466	272,024	219,747	166,309	1,358,820
State governments	924	1655	1334	1808	1462	1560	8743
Total	133,387	262,465	308,800	273,831	221,209	167,870	1,367,563

Capital expenditure over FY20 to FY25 is shown in Table 4.9.

Status of high value identified projects which are targeted by FY2023 is shown in Exhibit 4.12.

4.8 Conclusion

The investment plan of IR will ensure network expansion, decongestion, and safety in a big way and ensure availability of rake/seat on demand and better premium, high-speed goods, and passenger trains to sustain and augment its passenger and goods revenues. However, timely completion of capacity enhancement works and logistic hubs is a must to cater to the needs of the commuters/rail users and tap the future demand in various places. Promotion of private participation will hold key in fructification of IR's infrastructure development plans. Therefore, IR should facilitate public–private partnership to develop infrastructure using government as well as private resources.

Exhibit 4.1
Physical performance, productivity of Indian Railways

Year	Route kms	Running km	Total track km	Electrification in km	Railway stations in nos.	PKMS in billions	NTKM in billions	No. of originating passenger in millions	Tonnage originating in millions
1950–51	53,596	59,315	77,609	388	5976	66.517	37.56	1284	73.2
1960–61	56,247	63,602	83,706	748	6523	77.665	72.33	1594	119.8
1970–71	59,790	71,669	98,546	3706	7066	118.12	110.69	2431	167.9
1980–81	61,240	75,860	104,480	5345	7035	208.55	147.65	3613	195.9
1990–91	62,367	78,607	108,858	9968	7100	295.64	235.78	3858	318.4
2000–01	63,028	81,865	108,706	14,856	6843	457.02	312.37	4833	473.5
2010–11	64,173	87,114	114,037	19,607	7133	978.50	625.72	7651	921.73
2015–16	66,252	92,084	119,630	23,555	7216	1143.04	654.48	8107	1101.51
2016–17	66,918	93,902	121,407	25,367	7309	1149.84	620.18	8116	1106.15
2017–18	66,935	94,270	122,873	29,228	7318	1177.7	692.92	8283	1159.55
2018–19	67,415	95,981	123,542	34,319	7321	1157.17	738.52	8286	1221.48
2019–20	67,956	99,235	126,366	39,329	7325	1050.73	707.66	8086	1208.41
Analysis 2018–19	125.78%	162%	159%	8845.10%	122.51%	1739.66%	1966.24%	645.33%	1668.69%
2019–20	127%	167%	163%	10,136%	123%	1580%	1884%	630%	1651%

Source

1. Indian Railways Year Books from 1950–51 to 2019–20
2. Indian Railways Annual Report and Accounts from 1950–51 to 2019–20
3. Annual Statistical Statements of Indian Railways from 1950–51 to 2019–20

Exhibit 4.2
Throwforward into 2019–20 (Rs. in crore)

Plan head		Throwforward 2019–20
11	New lines (construction)	150,682
14	Gauge conversion	16,010
15	Doubling	201,944
16	Traffic facilities–yard remodelling and others	30,650
17	Computerization	2677
18	Railway research	418
29	Road safety works—level crossings	3177
30	Road safety works—road over/under bridges	40,305
31	Track renewals	84,146
32	Bridge works	6462
33	Signalling and telecommunications	93,391
35	Electrification projects	14,616
36	Other electrical works, including TRD	6629
41	Machinery and plant	2928
42	Workshops including production units	22,404
51	Staff quarters	2441
53	Passenger amenities	7848
64	Other specified works	7721
65	Training/HRD	2418
81	Metropolitan transport projects	101,741
		7,98,608
	*Excludes works costing less than Rs. 2.5 crore each	
	*Excludes plan heads rolling stock and investment in PSUs/JVs/SPVs and lease payments	
21	Rolling stock	2,81,300
		10,79,908

Exhibit 4.3
Growth in the number of locomotives and tractive efforts per loco

Year	Number of locomotives				Tractive effort per loco (in kg)	
	Steam	Diesel	Electric	Total	BG	M.G.
1950–51	8120	17	72	8209	12,801	7497
1960–61	10,312	181	131	10,624	14,733	8201

(continued)

(continued)

Year	Number of locomotives				Tractive effort per loco (in kg)	
	Steam	Diesel	Electric	Total	BG	M.G.
1970–71	9387	1169	602	11,158	17,303	9607
1980–81	7469	2403	1036	10,908	19,848	10,429
1990–91	2915	3759	1743	8417	24,088	12,438
2000–01	54	4702	2810	7566	29,203	18,537
2010–11	43	5137	4033	9213	34,380	18,304
2015–16	39	5869	5214	11,122	37,483	17,853
2016–17	39	6023	5399	11,461	37,808	17,746
2017–18	39	6086	5639	11,764	38,166	16,879
2018–19	39	6049	6059	12,147	39,413	16,226
2019–20	39	5898	6792	12,729	39,037	16,454

Source Indian Railways Year Book 2019–20

Exhibit 4.4
Growth in the number of Wagons and carrying capacity

Year	Total wagons on line (in units)	Covered	All gauges	
			Total no. of wagons* (000)	Total capacity (mill tonnes)
1950–51	205,596	58.9	195	4.14
1960–61	307,907	57.3	295	6.30
1970–71	383,990	53.4	368	9.35
1980–81	400,946	53.3	387	11.14
1990–91	346,102	49.1	335	11.50
2000–01	222,193	34.1	214	10.19
2010–11	229,987	26.6	217	12.18
2015–16	251,295	24.9	237	14.39
2016–17	277,992	24.0	264	15.99
2017–18	279,311	23.7	264	16.28
2018–19	289,185	23.4	275	16.95
2019–20	293,077	22.8	278	17.44

Source Indian Railways Year Book 2019–20
  *Excludes departmental service wagons and brake vans

Exhibit 4.5
Growth in the number of coaching stock on IR

Year	Passenger coaches						Other coaching vehicles	Total
	EMU coaches		Conventional coaches		DMU/DHMU			
	Number	Capacity	Number	Seating capacity	Number	Seating capacity	Number	Number
1950–51	460	87,986	13,109	854,678	–	–	6059	19,628
1960–61	846	150,854	20,178	1,280,797	–	–	7415	28,439
1970–71	1750	340,541	24,676	1,505,047	–	–	8719	35,145
1980–81	2625	500,607	27,478	1,695,127	–	–	8230	38,333
1990–91	3142	609,042	28,701	1,864,136	–	–	6668	38,511
2000–01	4526	859,701	33,258	2,372,729	142	13,884	4731	42,657
2010–11	7292	1,364,948	45,082	3,254,555	761	74,097	6500	59,635
2015–16	8805	1,578,868	53,171	3,794,954	1469	136,594	6704	70,149
2016–17	9125	1,646,880	53,668	3,969,607	1492	143,395	6699	70,984
2017–18	9556	1,748,490	54,080	3,957,263	1690	167,185	6499	71,825
2018–19	10,439	1,885,610	55,282	4,039,652	1876	164,976	6406	74,003
2019–20	11,360	2,072,843	57,083	4,205,915	1793	157,012	6372	76,608

Exhibit 4.6
Comparative position of train accidents

Year	Collision	Derailment	Level crossing accidents	Fire in trains	Misc accident	Total	Train accident per million train km
1970–71	59	697	122	13	–	891	1.91
1980–81	69	825	90	29	–	1013	1.96
1990–91	41	446	36	9	–	532	0.86
2000–01	20	344	83	15	2	464	0.64
2010–11	5	78	53	2	1	139	0.14
2017–18	3	53	13	3	0	72	0.06
2018–19	0	46	6	6	1	59	0.05
2019–20	5	40	1	7	1	54	0.05

Source
1. Indian Railways Year Books from 1970–71 to 2019–20
2. Indian Railways Annual Report and Accounts from 1970–71 to 2019–20
3. Annual Statistical Statements of Indian Railways from 1970–71 to 2019–20

Exhibit 4.7
Financial performance in Indian Railway

Year	Gross revenue Rs. in crores	Working expenses incl. depreciation, etc., and misc exp	Net revenue Rs. in crores	Capital at charge (Rs. in crore)	Total investment (Rs. in crore)	Dividend to general revenues and payment to states in place of tax on passenger fares	Excess/shortfall	Return on investment in %	Operating ratio in %
1950–51	263.30	215.74	47.56	827.0	855.2	32.51	15.05	5.75	81
1960–61	460.42	372.55	87.87	1520.9	1868.6	55.86	32.01	5.77	78.75
1970–71	1006.95	862.22	144	3330.3	4099.4	164.57	−19.84	4.35	84.13
1980–81	2703.48	2575.99	127.49	6096.3	7448.4	325.36	−197.87	2.09	96.07
1990–91	12,451.55	11,337.77	1113.78	16,125.8	22,200.5	938.11	175.67	6.91	91.97
2000–01	36,010.95	34,969.72	1070.23	43,051.88	63,341.01	307.64	763.59	2.49	98.34
2010–11	96,681.02	90,334.88	6346.14	143,220.57	231,615.25	4941.25	1404.89	4.43	94.59
2015–16	168,379.60	149,151.13	19,228.48	275,135.23	419,123.61	8772.51	10,505.97	7.0	90.50
2016–17	165,382.00	160,469.00	4913.00	302,457.78	471,776.39	–	4913.00	1.62	96.5
2017–18	178,929.64	177,264.03	1665.61	324,725.64	517,324.19	–	1665.61	0.51	98.40
2018–19	190,507.37	186,733.51	3773.86	348,601.77	573,641.66		3773.86	1.08	97.29
2019–20	174,694.69	173,105.07	1589.62	374,921.58	640,408.27	–	1589.62	0.29	98.36
Analysis 2018–19	723.5	865.5	79.3	Increase since 1950–51 453.4	Increase since 1950–51 748.8		250.8		

(continued)

(continued)

Year	Gross revenue Rs. in crores	Working expenses incl. depreciation, etc., and misc exp	Net revenue Rs. in crores	Capital at charge (Rs. in crore)	Total investment (Rs. in crore)	Dividend to general revenues and payment to states in place of tax on passenger fares	Excess/shortfall	Return on investment in %	Operating ratio in %
2019–20	663.5	802.4	33.4				105.6		

Source

1. Indian Railways Year Books from 1950–51 to 2019–20
2. Indian Railways Annual Report and Accounts from 1950–51 to 2019–20
3. Annual Statistical Statements of Indian Railways from 1950–51 to 2019–20
4. The operating ratio during 2019–20 would have been 114.19% had required level of appropriation to Pension Fund was made during the year
5. Dividend abrogated from 2016–17

Exhibit 4.8
Social service obligation on Indian Railways

S. No.	Description	2011–12	2012–13	2013–14	2014–15	2015–16	2016–17	2017–18	2018–19	2019–20
i.	Loss- on low-rated commodities	57	37	53	69	41	42	60	60	301
ii.	Concession in passenger fare	921	1045	1313	1423	1603	1670	1810	1995	2059
iii.	Losses on account of EMU suburban services	2814	3365	4027	4679	5125	5324	6184	6754	6938
iv.	Losses on uneconomic branch lines	1366	1466	1681	2056	1895	1855	2042	2342	2397
v.	Losses on strategic lines	733	904	954	1138	1225	1450	1797	1815	1604
vi.	Losses on parcel, luggage, postal and catering services, etc.	2807	2894	3346	3469	3808	3705	4729	5278	5793
vii.	Losses on account of pricing of fares below cost and other social service obligations	15,291	17,156	20,746	20,726	22,262	25,561	31,128	37,673	45,431
viii.	Total losses on coaching and goods services (i.–vii.)	23,989	26,868	32,120	33,560	35,959	39,607	47,751	55,917	64,523
ix.	Total social service obligation (viii.–vi.)	21,182	23,974	28,774	30,091	32,151	35,902	43,022	50,639	58,729
x.	Deduct staff welfare and law and order costs	6338	6785	7234	8213	8933	9968	10,664	12,326	13,187
xi.	Net social service obligations (ix.–x.)	14,844	17,188	21,540	21,878	23,218	25,934	32,358	38,314	45,542

Source Indian Railways Year Books from 2011–12 to 2019–20

Exhibit 4.9
Human resource and value added by human resource in Indian Railways

Year	Human resource in lakhs	Human resource cost Rs. in crores	Average wage per employee (Rs.)	Passenger earnings Rs. in crores	Freight earnings Rs. in crores	Gross revenue Rs. in cr	Average revenue earned per employee (Rs.)	Value added by human resource in IR (Rs.)
1950–51	9.14	113.8	1245	98.2	139.3	263.30	2881	1636
1960–61	11.57	205.2	1774	131.6	280.5	460.42	3979	2205
1970–71	13.74	459.9	3347	295.6	600.7	1006.95	7329	3982
1980–81	15.72	1316.7	8376	827.5	1550.9	2703.48	17,198	8822
1990–91	16.52	5166.3	31,273	3144.7	8247	12,451.55	75,373	44,100
2000–01	15.45	18,841.4	121,951	10,483	23,045.41	36,010.95	233,081	111,130
2010–11	13.32	51,776.57	388,713	25,705.64	60,687.05	96,681.02	725,833	337,120
2011–12	13.06	56,680.57	434,001	28,246.43	67,743.62	106,245.28	813,517	379,516
2015–16	13.30	93,001.24	699,257	44,283.00	109,208.00	168,379.60	1,266,012	566,755
2017–18	12.70	128,714.74	1,017,596	48,643.14	117,055.40	178,929.64	1,408,895	391,299
2018–19	12.28	135,171.13	1,097,370	51,067	127,433.00	190,507.37	1,551,363	453,993
2019–20	12.54	154,214.17	1,230,641	50,669	113,488	174,694.69	1,393,100	162,459
Analysis 2018–19	134.4%	118,779.6%	88,142.2%	52,003.1%	91,481.0%	72,353.7%	53,852.9%	27,754.5%
2019–20	137.2%	135,513.3%	98,846.7%	51,597.8%	81,470.2%	66,348.2%	48,359.0%	9931.8%

Source
1. Indian Railways Year Books from 1950–51 to 2019–20
2. Indian Railways Annual Report and Accounts from 1950–51 to 2019–20
3. Annual Statistical Statements of Indian Railways from 1950–51 to 2019–20

Exhibit 4.10
Modal share of traffic—road versus rail

Year	Road transport-freight movement (numbers in billion tonnes kilometres)	Road transport-freight movement (percentage of total freight movement)	Railways-freight movement (numbers in billion tonnes kilometres)	Railways-freight movement (percentage of total freight movement)	Road transport-passenger movement (numbers in billion passenger kilometres)	Road transport-passenger movement (percentage of total passenger movement)	Railways-passenger movement (numbers in billion passenger kilometres)	Railways-passenger movement percentage of total passenger movement)
1950–51	6.00	13.76	37.60	86.24	23.00	25.70	66.50	74.30
1960–61	14.00	16.22	72.30	83.78	80.90	51.01	77.70	48.99
1970–71	47.70	30.11	110.70	69.89	210.00	64.00	118.10	36.00
1980–81	90.90	38.10	147.70	61.90	541.80	72.20	208.60	27.80
1990–91	145.10	38.09	235.80	61.91	767.70	72.20	295.60	27.80
1999–2000	467.00	60.48	305.20	39.52	1831.60	80.96	430.70	19.04
2000–01	494.00	61.26	312.40	38.74	2075.50	81.95	457.00	18.05
2001–02	515.00	60.72	333.20	39.28	2413.10	83.10	490.90	16.90
2002–03	545.00	60.68	353.20	39.32	2814.70	84.53	515.00	15.47
2003–04	595.00	60.95	381.20	39.05	3070.20	85.01	541.20	14.99
2004–05	643.00	60.99	411.30	39.01	3469.30	85.77	575.70	14.23
2005–06	728.30	62.24	441.80	37.76	4251.70	87.35	615.60	12.65
2006–07	825.90	63.20	481.00	36.80	4657.00	87.02	694.80	12.98
2007–08	933.70	64.17	521.30	35.83	5482.00	87.68	770.00	12.32
2008–09	1021.60	64.95	551.40	35.05	6182.00	88.06	838.00	11.94
2009–10	1144.50	65.59	600.50	34.41	7192.00	88.84	903.40	11.16
2010–11	1287.30	67.29	625.70	32.71	8409.00	89.58	978.50	10.42
2011–12	1407.80	67.83	667.60	32.17	9478.00	90.06	1046.50	9.94
2012–13	1515.40	70.00	649.60	30.00	10,461.00	90.50	1098.10	9.50
2013–14	1650.80	71.26	665.80	28.74	11,730.00	91.14	1140.40	8.86
2014–15	1824.30	72.80	681.70	27.20	13,403.00	92.12	1147.20	7.88

(continued)

(continued)

Year	Road transport-freight movement (numbers in billion tonnes kilometres)	Road transport-freight movement (percentage of total freight movement)	Railways-freight movement (numbers in billion tonnes kilometres)	Railways-freight movement (percentage of total freight movement)	Road transport-passenger movement (numbers in billion passenger kilometres)	Road transport-passenger movement (percentage of total passenger movement)	Railways-passenger movement (numbers in billion passenger kilometres)	Railways-passenger movement (percentage of total passenger movement)
2015–16	2026.10	75.58	654.50	24.42	15,415.00	93.10	1143.00	6.90

Source

1. Indian Railways Year Books from 1970–71 to 2017–18
2. Indian Railways Annual Report and Accounts from 1970–71 to 2017–18
3. Annual Statistical Statements of Indian Railways from 1970–71 to 2017–18
4. Yearbook 2011–12 to 2015–16 of Ministry of Road Transport

Exhibit 4.11
Infrastructure projects identified under Vision 2024

Doubling

- 58 Super critical projects of total length 3750 km (balance length 1471 km), total cost Rs. 36,700 crore, throwforward Rs. 8100 crore have been planned to complete by Dec' 2021 except two projects completed by Mar' 2022.
- 68 critical projects of total length 6918 km (balance length 6075 km), costing Rs. 69,900 crore, throwforward of Rs. 45,400 crore will be completed by Mar' 2024.
- Additional 20 coal projects of total length 2350 km (balance length 1508 km), costing Rs. 29,800 crore, throwforward Rs. 15,000 crore are targeted for completion by Mar' 2024 (31 coal projects and 10 port connectivity projects are covered in super critical and critical projects). The coal projects include seven new lines of total length 619 km (balance length 377 km) costing Rs. 12,903 crore (throwforward of Rs. 4639 cr) and 1 gauge conversion project of length 285 km (balance length 24 km) costing Rs. 2005 crore (throwforward of Rs. 288 crore).
- In addition, 48 projects on HDN and HUN routes of total length 3700 km (balance length 3444 km), costing Rs. 42,800 crore, throwforward of Rs. 34,463 crore will be completed by Mar' 2024.

New Line: Focus Will Be to Complete National Projects

- 4 national projects falling in NE region of total length 294 km (balance length 282 km) costing of around Rs. 27,500 crore and throwforward Rs. 12,094 crore will be completed by Mar' 2023.
- Udhampur-Srinagar-Baramula Rail Link (USBRL) Project of total 272 km length (balance length 111 km), costing Rs. 28,000 crore and throwforward of Rs. 9949 crore by Dec' 2022.
- 2 projects (Rishikesh-Karanprayag and Bhanupalli-Bilaspur) of total length 188 km costing around Rs. 23,000 crore throwforward of Rs. 20,718 crore will be completed by Mar' 2025.

Gauge Conversion

- 30 gauge conversion projects of total length 3913 km (balance length 2292 km) costing around Rs. 31,200 crore and throwforward of Rs 15,053 crore 24 March will complete a total 2146 km gauge conversion.

Traffic Facility

- 105 traffic facilities work costing Rs. 10,150 crore with throwforward of Rs. 6782 crore will be completed by Mar' 2024.
- 146 railway electrification projects of balance length 23,800 km, throwforward of Rs. 20,209 crore will be achieved by Dec' 2023.

Signalling and Telecommunication Work

- Signalling and telecommunication work including the provision of TCAS, automatic signalling, LTE and CTC on HDN and freight dense HUN routes, elimination of electro-mechanical signalling, replacement of overaged signalling gears, interlocking of LC gates and other miscellaneous works costing Rs. 21,014 crore will be completed by Mar' 2024.

Raising of Sectional Speed

- Works for raising sectional speed to 160 kmph on New Delhi-Mumbai (incl. Vadodara-Ahmedabad) and New Delhi-Howrah (incl. Kanpur-Lucknow) routes costing Rs. 6804 crore and Rs 6684 crore, respectively, will be completed by August 2023.
- Infrastructural works for raising the sectional speed to 130 kmph on balance golden quadrilateral and diagonal routes will be completed and will increase 130 kmph by December' 2021.

Elimination of Manned-Level Crossings from GQ-GD Routes

- To enhance safety and operational speed on golden quadrilateral and golden diagonal, manned-level crossings (MLCs) should be removed on priority.
- As of 01.04.2020, 2429 MLC existed on GQ-GD routes. As per the survey done by Zonal Railways, 1186 LCs will be replaced by ROBs, 1223 by RUBs, and 20 LCs by diversion at a total approximately Cost of Rs. 41,200 crore work is targeted to be completed by March 2025.

Exhibit 4.12
Current status of high value identified projects targeted for completion by FY23

Project	Project initiation date	Estimated COD (updated COD, if any)	TPC (Rs. crore)	Actual expenditure till FY20	Projected expenditure FY21	Actual expenditure FY21 (Aug'20)	Remarks
Western dedicated freight corridor (WDFC) from JNPT to Rewari (1504 km)	Feb'2008	Dec-21 (Jun-22)	60,933	35,055	10,300	2280	The work is planned to be completed by Jun'2022
Eastern dedicated freight corridor (EDFC) from Ludhiana to Sonnagar (1318 km)	Feb'2008	Dec-21 (Jun-22)	33,729	25,452	6956	1952	The work is planned to be completed by Jun'2022
Udhampur-Srinagar-Baramulla Rail Link Project. The project length is 272 km, out of which 161 km has been commissioned	98–99	Jun-22 (Dec-22)	27,949	21,098	2658	403	The work is planned to be completed by Dec'2022
Jiribam-Imphal (NP) (110.625 km). It is a capital connectivity project of Manipur	October'2003	Dec-22 (Mar-23)	12,264	10,094	800	307	Frequent law and order issue hampers the progress
East–West metro corridor between Howrah Maidan and Salt Lake	Jun'2008	Mar-22	8575	6596	1533	268	Outlay for 2020–21 is 905 Cr
Bairabi-Sairangnew New Line Project in North East Region connecting state capital Mizoram (51.38 km)	Mar'2009	Mar-22 (Mar'23)	5521	3588	535	104	Work is planned to be completed by Mar'23
CEWRL (Gewra Road to Pendra Road BG DL and SL from Urga to Kusumunda)—NL in Chhattisgarh State	June'2013	Mar-22 (July'23)	4970	650	2000	148	Achieved financial closure in July'2020

References

1. Bryman, A. (2012). *Social research methods* (4th ed.).
2. Choong, K. K. (2013). Understanding the features of performance measurement system: A literature review. *Measuring Business Excellence, 17*, 102–121. https://doi.org/10.1108/MBE-05-2012-0031
3. Shepherd, J., Harden, A., Rees, R., Brunton, G., Garcia, J., Oliver, S., & Oakley, A. (2006). Young people and healthy eating: A systematic review of research on barriers and facilitators. *Health Education Research, 21*, 239–257. https://doi.org/10.1093/her/cyh060
4. Matthews, R. L., & Marzec, P. E. (2012). Social capital, a theory for operations management: A systematic review of the evidence. *International Journal of Production Research, 50*, 7081–7099. https://doi.org/10.1080/00207543.2011.617395
5. Pino, M.-T., Skinner, J. S., Park, E.-J., Jeknić, Z., Hayes, P. M., Thomashow, M. F., & Chen, T. H. H. (2007). Use of a stress inducible promoter to drive ectopic AtCBF expression improves potato freezing tolerance while minimizing negative effects on tuber yield. *Plant Biotechnology Journal, 5*, 591–604. https://doi.org/10.1111/j.1467-7652.2007.00269.x
6. Mark Petticrew, H. R. (2006). Why do we need systematic reviews? In: *Systematic reviews on social science* (pp. 1–26). Wiley. https://doi.org/10.1002/9780470754887.ch1
7. Prasad, C., & Jamuar, S. S. (2021). Optimising indian railways infrastructure by AI. *Journal of Mobile Multimedia, 17*, 157–174. https://doi.org/10.13052/jmm1550-4646.17138
8. Bhatia, V., & Sharma, S. (2021). Expense based performance analysis and resource rationalization: Case of Indian Railways. *Socio-Economic Planning Sciences, 76*, 100975. https://doi.org/10.1016/j.seps.2020.100975
9. Narayanaswami, S. (2018). Digital social media: Enabling performance quality of Indian Railway services. *Journal of Public Affairs, 18*, e1849. https://doi.org/10.1002/pa.1849
10. Monica, M., & Sharma, S. (2012). Financial turnaround of Indian Railways (A). *Emerald Emerging Markets Case Studies, 2*, 1–11. https://doi.org/10.1108/20450621211289485
11. Singhania, M., & Sharma, S. (2012). Financial turnaround of Indian Railways (B). *Emerald Emerging Markets Case Studies, 2*, 1–7. https://doi.org/10.1108/20450621211289494
12. Banerji, A. K. (2021). *Dedicated freight corridors will boost rail productivity*. Accessed August 29, 2021. https://trid.trb.org/view/792964
13. Singhania, M., & Sharma, S. (2013). Profitability and leverage analysis of indian railways: Impact of cost based indexation. *SSRN Electronic Journal*. https://doi.org/10.2139/ssrn.2312660

Part II
How Power Sector is Managed in India?

Chapter 5
Power Sector Infrastructure Management: Issues and Challenges

Anil Kumar and Avishek Ghosal

Abstract In the last decade, we observed that India had taken significant steps in reforming the power sector. Power sector is in transition phase. Shifting the focus from conventional resources to non-conventional resources, bringing in reforms to revive power distribution sector and enhancing competition in power market are some of the highlights. The reforms like National Solar Mission (NSM), Deen Dayal Upadhaya Grameen Jyoti Yojna (DDUGJY), National Smart Grid Mission (NSGM), Real Time Market (RTM) and Green term Ahead Market (GTAM) in power-trading, Ujjwal Discom Assurance Yojna (UDAY) and Integrated Power Development Scheme (IPDS) are some of the highlights. The generation sector had seen a significant growth to installed capacity of 375 GW. Still some barriers and challenges remain which are hampering and creating headache for policymakers to revive the power sector in India. The issues are both technical as well as commercial in nature. Some of them are also latent in nature. In addition, all these are impacting the sustainable economic growth of India. To resolve the issues and challenges, the reforms have to pass through the regulatory, technical and economic viability check so that it can be implemented. Through this paper, we tried to identify and analyze the bottlenecks and challenges through systematic literature review analysis. This paper also suggests some probable steps and framework for the policymakers for reforming power sector of India. This reforms can ensure a secured, affordable and reliable sustainable energy future of India.

5.1 Introduction

Energy is an important resource which feeds to economic growth of a country. With economic growth of India we observed for the last two decades, the country's appetite for energy also increased [1–3]. Same can be observed for India's power sector. At the start of the millennium with completion of IX-Five year plan in year 2002, our power generation installed capacity stood at 105,046 MW, transmission line of 304,258

A. Kumar · A. Ghosal (✉)
University of Petroleum and Energy Studies, Dehradun, India
e-mail: aghosal@ddn.upes.ac.in

circuit km, peak demand of 78,441 MW and per capita electricity consumption stood at mere 556 kWh as per Central Electricity Authority (CEA) report [4]. At present, our power generation installed capacity stands at 384,000 MW, transmission line of 725,200 circuit km, peak demand of 213,244 MW and per capita electricity consumption stood at mere 1208 kWh [4]. We can say from above data the power sector infrastructure has seen a growth by more than 2 times.

The growth in power sector can be said was accelerated by bringing in change in policy and regulatory initiative starting by bring in new Electricity Act of 2003. The act of 2003 also bought in competition by creating trading platform, open access policy, etc., accelerated the growth in India's power sector. In the last decade, we also saw a transitional shift from dependence on conventional resources to nonconventional resources like solar and wind resources. This transition was part of India's objective of achieving sustainable development. At 2007, the installed renewable energy stood at mere 7760 MW, and by end of 2020, renewable energy installed capacity stands at 91,154 MW [4]. India also surpassed Japan and Russia in becoming world third largest producer of electricity during this period [1]. After 2003, policymakers bought in other reforms through policy and regulations to sustain the growth that is required in the country's ever increasing energy appetite. The reforms like National Solar Mission (NSM), Deen Dayal Upadhaya Grameen Jyoti Yojna (DDUGJY), National Smart Grid Mission (NSGM), Real Time Market (RTM) and Green term Ahead Market (GTAM) in power-trading, Ujjwal Discom Assurance Yojna (UDAY) and Integrated Power Development Scheme (IPDS) are some of the highlights. But still lots of challenges and issues can be observed in the country's power sector infrastructure and services which is hindering the growth of the sector.

Through this research paper, we tried to highlight the prevalent issues and challenges existing in the power sector. The Indian power sector might have witnessed significant changes in the last two decades. We can still observe power shortages in some of India especially during peak period. Underperformance in operation segment and unjustified fiscal situation of distribution companies (DISCOMs) pressed a number of policy and regulatory changes [2]. These sort of incentive and reform in operational and governing attributes are likely to resolve the technical and fiscal issues that the Indian power sector are facing. This paper gives brief explanation of Indian power sector. In addition, the issues that have driven various footsteps undertaken by the central and the state governments for the progress of the sector and the key policy and regulatory changes in the power sector commenced since 2000 has been highlighted.

5.2 Literature Review

Electricity is a key force of economic growth for any country [1, 5]. The expansion of the Indian economy would depend heavily on the availability of quality power infrastructure [2]. India's per capita electricity consumption might have doubled in the last decade, but still it is less than world average of 3131 kWh. If India compares itself with BRICS country also its per capita consumption is less than China and Brazil.

India was able to reduce the peak load deficit from 7.85 to 0.51% [4]. But still some of parts of India is facing a peak load deficit of more than 5% like Union Territory (UT) of Ladakh, UT of Jammu & Kashmir and in all North Eastern States [6].

The cost of purchase of electricity and efficiency in operation of electricity is a key factor that determines the profitability of a distribution utility [7]. Cumulatively the factors determine the cost of supply [7, 8]. It has been observed that cost of purchasing electricity from generation companies constitutes 65–75% of total cost of supply of power for a distribution utility [7, 8]. It has been observed that there was an increase by 28–30% in the cost of purchase of electricity [2, 7, 8].

We all know that for a continuous economic growth a proper functional electricity sector is essential. For high rate industrialization and also for rural development electricity supply of electricity cannot be poor quality or unreliable [9].

Developing countries always turn to multilateral and bilateral financing organizations for funding domestic development projects through loans with some conditions [9, 10]. This takes place as developing countries have limited financial capacity. The loans also bring in reforms as part of condition stated by the multilateral financing organizations.

Political economy of market reform is an important factor that effects the outcome of any reform. Similarly can be observed for power sector in India also. Some obstacles and challenges that can be identified as a bottleneck for reforms are like electoral populism, partisan cleavages, interest groups, etc. One group of researchers states that that those who are profited from any reforms become more powerful and they build up pressure on policymakers and administrators [9]. Many also argue that reforms sometimes can undermine the office-seeking politicians. As stated by Przeworski that before achieving expected benefits out of a reform there have to be a sustain political and economic instability [9, 11].

Deploying renewable energy can be observed in many countries to reduce carbon footprint and resolving the effect of climate change [12]. Power supplied to consumers should be uninterrupted, stable, reliable and affordable. In India, capacity addition of renewable energy has helped in reducing power deficiency [1–3]. After 2010, India had picked its growth in capacity addition of renewable energy. But due to COVID-19 pandemic somewhat the growth rate got decreased [13, 19].

5.3 Objective of the Study

- To understand the various challenges and risk in power sector
- To suggest solution and remedies to the various problems in power sector in India.

5.4 Methodology

A systematic review method of literature has been applied for the present study to extract the factors that are creating hindrance in the process of development of a region, with focus on power sector development. Arksey and O'Malley's systematic review framework was considered as reference for the review process [1]. The central research question of the paper is "what are the issues and challenges to power sector in India". For review of the literature, particular journals were not predetermined; instead, we used the following sources: EconLit (EBSCO), JSTOR, Science Direct and Wiley Online Library. With respect to research question, journals and articles were searched from electronic databases with systematic review procedure. Key terms like "Power sector reforms", "Power Business", "Renewable Energy" and "Distribution sector" were used to search for identifying key journals and articles. Terms like "social", "ecology or environment" and "economic" were also used in the reviewing process. Government reports were also considered for this study.

5.5 Power Sector in India: Issues and Challenges

As the Indian power sector is embarking on increasing the generation and transmission capacities, key challenges lie ahead which also resulted in the historical underperformance.

1. Underperformance: India has historically failed to meet its power sector targets by a significant margin, and with tremendous opportunities ahead, the power sector continues to be affected by the shortfall both on generation as well as transmission side. For example, for the current installed capacity of around 375 GW, the inter-regional transmission capacity is only about 103 GW (27% of the installed capacity). The various proposals in generation and transmission are currently under different implementation stages. However, the power sector in India has been plagued with a set of problems for meeting the planned targets. Although measures have been defined by the policymakers and stakeholders in a sense of complacency that the issues will indeed be resolved and India will plug the supply deficit of power to resolve the same, but looking at the past record, it can be estimated that the resolution measures may not be implemented. The biggest indicator of a poor track record is the inability to meet targets on the power generation capacity additions [1, 2, 4, 5, 14].
2. Land Acquisition and Environment Clearance: Land acquisition poses an increasingly significant challenge in the Indian electricity sector. Power plants and utilities face major constraints and delays regarding the availability of land and obtaining the requisite environment and other clearances for the projects. The new bill relating to land acquisition has continued to face political opposition. While it provides for acquisition by project development agencies to the extent of 70% of the land required for a project, with the balance to be obtained

by the government. In addition, it has been reported that in some cases, even after land owners were asked to sell and handover their land in "Public Interest", the project was not completed for several years due to other delays, a fact that eroded the credibility of both the industry and the government. Consequently, there is a significant mismatch of expectations from the project affected persons (PAP). Stakeholders or other land owners may collectively object of the project execution. In such cases, it is essential to proactively manage the environment and stakeholders' expectations [5, 6, 12].

3. Scheduling of Electricity through Transmission Lines: Substantial development in construction of transmission lines is likely to be required to meet the additional installed generation capacity especially for renewable energy. Substantial portion of this development is required in the Himalayan belt, which have difficult terrain reducing the margin of error for project execution. Additional transmission capacity is required to evacuate power from surplus regions to supply the power deficit states [1, 2, 5, 15].

4. Reform Implementation: Few states like Delhi, Gujarat, and West Bengal can be considered as the model states, where the power sector reforms were successfully implemented. Furthermore, the new policies enabled considerable improvements in the performance of the power sector. In contrast, reform efforts fell flat in six states, including Bihar, Jharkhand, Kerala, Tamil Nadu, Uttar Pradesh and Uttarakhand. Although some reforms were implemented, they were limited and yielded disappointing substantive results [8, 9, 16]. Technical and financial considerations were at the forefront in almost all of the states, and power sector reforms were a response to major difficulties and weak performance of electric utilities. The partially successful cases of Odisha and Maharashtra, for instance, both aimed to decrease their large T&D losses through restructuring. Maharashtra's T&D loss prior to reform was 35% in 2004, and this figure decreased to 15% in 2014. In contrast, Odisha continues to have one of the highest T&D and AT&C losses despite having a revised tariff structure. Other studies suggest that electricity pricing and collection may help explain these different rates of improvement. In Maharashtra, several measures have been taken to improve bill collection and reduce theft such as electronic payment systems that has resulted in improvements in T&D and AT&C losses [9, 17].

5. Partisan Cleavages: The most common problem was labor union resistance, followed by agricultural lobby activism. Partisan conflicts were relatively uncommon and took place in only in few states. Labor union opposition was found in case of Delhi, and representatives of Delhi Vidyut Board's (DVB) employees reached an agreement with the Delhi government on reform. On the other hand, unions were by far the most common obstacle in cases of failure of reform implementation. One example was when State Electricity Board (SEB) employees and engineers organized strikes in protest of the state government's unbundling and privatization plans and they had a large impact on delaying negotiations. For example, in Bihar unbundling occurred in 2013 after much postponement due to union protests. Thus, labor union association is strongly associated with the tendency of reforms to fail. Farming lobbies also played a

key role in many cases of failure. In the large states of Tamil Nadu and Uttar Pradesh, effective mobilization by farmers was associated with failure. Since farmers in these states often enjoyed free or nearly free electricity through subsidies, organized farmers' lobbies mobilized against the threat of increased agricultural electricity prices and threatened to punish the government for reforms in elections. For instance, in Uttar Pradesh, the agricultural sector received high subsidies, and farmers have held protests across the state against tariff hikes [9, 11, 16].

6. Challenges for Large-Scale Solar Developers in India: Current policy and regulatory environment is creating hurdle for solar power developers as there are numbers of constraints and gaps. This section describes five most critical barriers, technological barriers, policy and regulatory barriers, financing barriers, transparency and accountability and infrastructure barriers, which affect the transition to solar power development in India [3, 5, 13, 18]. India is trying to increase manufacturing facilities and R&D facilities of solar in last ten years but, due to lack of policies, lack of awareness, poor quality, lack of financial support have undermined this effort. Almost every developed country and most of the developing countries are investing in R&D in solar energy and setting up large manufacturing facilities for the growth and development of the solar power sector. According to the Ministry of New and Renewable Energy (MNRE), in the year 2016, India's cell and module manufacturing capacity now stand at 1212 MW and 5620 MW, respectively, whereas countries like USA, China, Germany, Malaysia, etc., are capable of multi-giga watt production.

7. Transparency and Accountability: Transparency and accountability are key principles for good governance, while good governance is pre-condition to achieving sustainable development. In 2005, India was ranked 76th in Transparency International's Corruption Perceptions Index, and the main causes of corruption are excessive regulations, complicated taxes and licensing systems, numerous government departments, each with opaque bureaucracy and discretionary powers, monopoly by government controlled institutions on certain goods and services delivery and the lack of transparent laws and processes. Major source of corruption in India occurs in spending money in government schemes and programs, and therefore, the growth of large-scale solar power is moving at a very slow rate. About 28% of the developers have agreed that directly or indirectly they are forced to pay bribe starting from the contract stage till the completion of the project. They are also in the favor that there is no transparency in issuing contracts to the developers and in land acquisition. Even the disbursal of the National Clean Energy Fund (NCEF) and its operational modalities has some issues with respect to transparency. The NCEF has been operating since financial year 2010–11 and has been conceived of as a separate non-lapsable corpus to support research and innovative projects in the field of clean energy technology of which, major portion will be utilized in solar PV. The fund is created through the levy of a clean energy cess on both domestic and imported coal (Rs. 50 per tonne). It is being administered by the Ministry of Finance. The fund has been largely criticized for inconsistencies between the

stated objectives, operational guidelines, and final approval of the projects. It has been found that instead of funding cutting-edge R&D on clean technology, it is being used to cover the budgetary shortfalls for projects implemented by various ministries and departments. Another example of poor transparency has been identified in the state of Rajasthan where under National Solar Mission (NSM) seven associate companies of a private company named as LANCO were granted commissioning certificates of a 5 MW solar power project by the Rajasthan Renewable Energy Corporation Limited (RRECL). However, it has been found that the certificate was granted even before the plant was completed, in direct violation of the guidelines of NSM [3, 5, 12, 17].
8. Cost of Supply of Electricity: Electricity Distribution Companies (DISCOMs) are important part of electricity value chain. They are critical part in delivery of last mile supply to the end-consumer. The DISCOMs action also reflects the government's vision of reflecting real and judicious cost in electricity tariff charged from end-consumers both domestic and commercial. For tariff fixation, determination of cost of supply is an important task. It reflects the cost imposed to a consumer to avail supply from the DISCOMs. To estimate the cost of supply, simplified/average approach is being followed by most of the DISCOMs. To discover correct price, better methodology is required. These require a large volume of voltage and consumer-category-wise data. The availability of which is one of the biggest challenges faced by Indian DISCOMs [3, 5, 8, 12, 14].

5.6 Remedies and Solutions

The deficit in power accessibility and availability in India is a major weakness to the growth of the country's economy. In the present framework, linking the deficiency in demand and supply of electricity has become critical, and so, projects are being undertaken in various segment of the power value chain, generation, transmission and distribution. In addition, India has not seen such a large scale of projects before. So, there is a requirement for analyzing and improving project performance abilities to help ensure targets are completed in time.

These demands employing an inclusive project management framework to resolve the key challenges and issues of the power sector projects and to be completed as per the scheduled targets. Analysis of past data also reflects existence of a weak project management framework. These frameworks do not assess all the key factors that lead to various issues and challenges. Initially, we highlighted that this paper is to highlight the opportunities and challenges of the electricity sector. Thus, we can say that project management solutions and remedies that are required to address the challenges mentioned [10, 12, 15, 17].

Through our reviewing process, we had identified that one type of technology can solve a specific challenge. It appeared instinctive from an analytical standpoint. But business and policymakers do not sufficiently recognize solution technologies as alternatives for answering certain challenges. Not incorporating developments of other solution technologies can reduce the viability of single technology. Such misinterpretations can contribute to temporary market price decrease. The example of such misinterpretation is the case of Germany's balancing power markets. These changes in the institutional framework and simultaneous development of storage, demand response, and improved variable renewable energy forecasting has led to a substantial decline in market size and prices over the last few years[3, 5, 17].

References

1. Kumar, D., & Hinduja, D. K. P. B. (2015). Issues and challenges in electricity sector in India. *The Business & Management Review, 5*(4), 29–30.
2. Singh, A. (2006). Power sector reform in India: Current issues and prospects. *Energy Policy, 34*(16), 2480–2490.
3. Kumar, P., Rathore, S., Rathore, S., & Singh, R. P. (2022). Solar power utility sector in India: Challenges and opportunities. *Renewable and Sustainable Energy Reviews, 81*, 2703–2713.
4. CEA. (2020) *Growth of electricity sector in India from 1947-2020.*
5. Rathore, P. K. S., Rathore, S., Pratap Singh, R., & Agnihotri, S. (2018). Solar power utility sector in India: Challenges and opportunities. *Renewable and Sustainable Energy Reviews, 81*, 2703–2713 (March 2017).
6. Authority, C. E. (2021). *Government of India Ministry of power executive summary on power sector.*
7. CERC. (2019). *Consultation paper on terms and conditions of tariff for tariff period.*
8. Pachouri, R., Raparthi, B., & Sharma, A. (2020). *Cost of supply of electricity a systemic review and way forward for Indian DISCOMs.*
9. Yo Cheng, C., Lee, Y. J. J., Murray, G., Noh, Y., Urpelainen, J., & Van Horn, J. (2020). Vested interests: Examining the political obstacles to power sector reform in twenty Indian states. *Energy Research & Social Science, 70*, 101766 (March).
10. Collier, P., Guillaumont, P., Guillaumont, S., & Gunning, J. W. (1997). Redesigning conditionality. *World Development, 25*(9), 1399–1407.
11. Przeworski, A. (1991). *Democracy and the market: Political and economic reforms in Eastern Europe and Latin America.* Cambridge University Press.
12. Sinsel, S. R., Riemke, R. L., & Hoffmann, V. H. (2020). Challenges and solution technologies for the integration of variable renewable energy sources—A review. *Renewable Energy, 145*, 2271–2285.
13. Madurai Elavarasan, R., et al. (2020). COVID-19: Impact analysis and recommendations for power sector operation. *Applied Energy, 279*, 115739 (May).
14. CEA. (2018). *National electricity planning.*
15. CEA. (2020). *Government of India Ministry of Power.*
16. Karmacharya, S. B. (2012). Lessons to be learned from the experience of electricity reforms in India. *Hydro Nepal Journal of Water, Energy and Environment*, 29–36.
17. Totare, N. P., & Pandit, S. (2010). Power sector reform in Maharashtra, India. *Energy Policy, 38*(11), 7082–7092.
18. Hayashi, D. (2020). Harnessing innovation policy for industrial decarbonization: Capabilities and manufacturing in the wind and solar power sectors of China and India. *Energy Research & Social Science, 70*, 101644 (June).
19. Kumar, A., Ghosal, A., & Rab, M. T. (2021). Proposed Amendment of Electricity Distribution in India: A review. *Water and Energy International, 63*(10), 28–34.

Chapter 6
Renewable Energy Management: An Analysis of the Status Quo

Nikita Das

Abstract India has witnessed a significant transition in its electricity mix since the landmark announcement in 2015 targeting the achievement of 175 GW of Renewable Energy (RE) by 2022. Ever since, RE installed capacity has exponentially risen. Given this rise in RE integration, there is a need to understand RE infrastructure management through the facets of technological patronage, favourable economics and conducive policy and regulatory framings. This chapter underscores the governmental, institutional and regulatory infrastructures in place for RE management, and in doing so, it aids in identifying the various issues and challenges that pertain to such management and proposes suggestions to circumvent the same. While RE cannot be deemed as a silver bullet at the outset, acknowledging the issues it is plagued with can pave the way for establishment of meaningful infrastructure and its effective management in the long run.

Keywords Renewable energy · Infrastructure · Management · Technological interventions · Economics · Regulatory frameworks · Just energy transition

6.1 Introduction

The United Nations recently announced India as one of the Global Champions in energy transitions for a High-Level Dialogue on Energy 2021. This is reflective of the considerable steps India has taken to be regarded as a force to reckon with on renewable energy (RE) infrastructure in the last decade. Since the humble beginnings of RE capacity addition in the 2000s to the ambitious target of 175 GW by 2022 and 450 GW by 2030, RE integration has come a long way in India. In the G-20 Summit organised on 21 November 2020, the Indian Prime Minister, Narendra Modi shared with confidence that these targets will be achieved with ease. The period until now has witnessed considerable changes in technological upgradations, innovations, economic models and the regulatory frameworks in place. The sector itself

N. Das (✉)
Department of Anthropology, University at Buffalo, SUNY, Buffalo, USA
e-mail: ndas@buffalo.edu

has undergone transformations on account of the increased policymaking impetus for solar and wind-based technologies for power generation. Subsequently as Fig. 6.1 highlights, India has etched a trailblazing pathway for itself over the past decade through the central and state policy schemes that have aided in the steady rise of RE installed capacity for power generation.

Figure 6.2 shows that the trend has held even in 2021 as the component of RE in the total installed capacity in the country has increased from 18% in 2018 to 24% at present.

Central government policies have moved in tandem with the market indicators as RE achieved grid parity in comparison with the fossil fuel-based energy sources

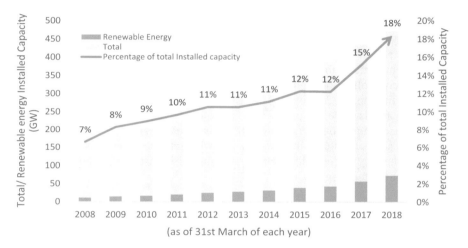

Fig. 6.1 Yearly renewable energy installed capacity from 2008 to 2018. *Source* Analysis by author with data collated from yearly 'Energy Statistics' published by the Ministry of Statistics and Programme Implementation

Fig. 6.2 Installed capacity mix as of 28.02.2021 (in GW). *Source* Ministry of Power [1]

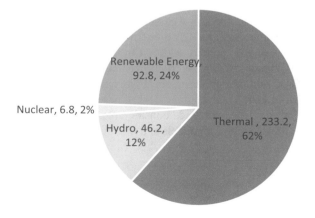

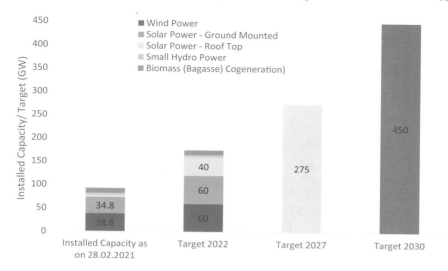

Fig. 6.3 Renewable energy installed capacity as of 28.02.2021 and targets according to government announcements and strategic blueprints. *Source* Data collated by author from various reports [2–4]

for electricity. However, this does not imply that the sector has achieved these feats devoid of any obstacles. In fact, the sector continues to be plagued with operational and technological issues. As Fig. 6.3 elucidates, the installed capacity in 2021 at a little less than 100 GW is still considerably short of the 2022 targets. As technology, conducive economics and supportive institutions and regulatory frameworks work like spokes of a wheel that can keep the barrel of energy transition enroute its long journey ahead, it is important to acknowledge these three spokes of RE infrastructure in India. Thus, the infrastructural advances that India has made in alignment with the power sector must be analysed in cognisance with the socio-political economy. This chapter makes this endeavour possible by providing a holistic overview of the infrastructures in place in the RE sector that have enabled the commencement of the Indian energy transition. The subsequent sections delve deeper into the various facets of the energy transition of India.

6.2 Technological Interventions and Patronage

Renewable energy-based technological development in the country has come in the form of new innovative solutions that strive for greater RE penetration with more ease. Technological interventions have been seen to actuate social and economic changes. This holds true for RE-based technologies as well. These interventions work in tandem with policies and schemes introduced by governments in place. However, the challenge of operationalising these technologies successfully is heavily dependent on institutional frameworks in place and the political economy in general. The latter is a major blindside as a lack of understanding of the political economy, and therefore, inability to pre-empt challenges prevents streamlined and seamless implementation of these technologies. This results in failure in many of these technologies gaining traction.

6.2.1 Solar-Based Applications in Agriculture

Of all the forms of RE-based technologies for agricultural purposes, solar has been deemed the most promising. Such technologies can be used for various agricultural purposes. Solar power in farming has been forged to be an ideal solution for irrigation in developing countries [5]. Agricultural electric pumps that are used for irrigation can be powered by solar plants or can be replaced with solar-powered pumps [6]. Stand-alone solar pumps are the primary substitute for agricultural electric and diesel pumps. Such technologies are much more environment-friendly, unlike diesel pumps as they use no fuel [7].

The Indian central and state governments have come forward with similar grass root-level innovations. The Prime Minister KUSUM scheme was launched by the central government in early 2019 that promoted the installation of solar pumps and grid connected solar plants by farmers. In the first instalment, the government pushed for individual solar plants of up to 2 MW to be instated for agricultural use and substitute the use of agricultural pumps [8]. This policy scheme would allay two issues. First, it could reduce the dependence on fossil fuel-based sources of energy. Second, it could decrease the electricity distribution companies' (DISCOMs') subsidy burden given that farmers become more self-reliant for agricultural use of electricity. If not for subsidised electricity tariffs for using irrigation pumps, agricultural costs would be substantially high [9]. Solar-based irrigation systems can circumvent these subsidy-related issues to a great extent [10].

Solarisation of agriculture was also encouraged by states like West Bengal and Haryana who introduced subsidies as high as 75% and 90%, respectively, for purchase of solar pumps [11]. Andhra Pradesh has also proposed a plan to install 10 GW of solar plants to supply free day time power to farmers [12]. A landmark model was introduced in Gujarat, namely the Solar Power as a Remunerative Crop or 'SPaRC' model. In this model, India's first solar irrigation cooperative was set up in 2016 so as to install solar pumps. Farmers sought to go beyond crop diversification with the emergence of treatment of solar as remunerative by way of sale of excess solar energy generated and sale of excess water extracted [13]. The farmers in the cooperative signed a power purchase agreement with the electricity distribution company for the sale of excess solar energy from solar pumps at a predetermined tariff. By signing the agreement, the farmers gave up the power to an agricultural connection for the next twenty-five years (equivalent to the life of a solar panel). Policy impetus in the form of 80–90% subsidy also ensured that interested farmers could opt for solar pumps [14]. The price the farmers received for the sale of electricity was ascertained to be higher than the usual RE tariffs, as the solar pumps were replacing diesel pumps and future subsidy requirements of the farmers [15]. The 'SPaRC' model paved the way for a promising alternative decentralised energy system in the form of solar pump systems.

While the 'SPaRC' model is unique and promising, it was critiqued by stakeholders within the electricity and water space. The project was evaluated within eighteen months of its initiation when the ideal period to evaluate a project's performance is after three years [16]. Additionally, the upfront installation costs were grossly underestimated [17]. There is also uncertainty regarding the shelf life of solar pumps that farmers buy with their hard-earned money. Finally, the impact of the dependence of farmers on a feed-in-tariff, on already cash strapped DISCOMs, is an unknown [18]. Although energy cooperatives can play a crucial role in uplifting small farmer livelihoods, they are still at a nascent stage in the country. This implies that preclusion of small and marginal farmers may be common. Hence, circumventing the issues of accessibility would require considerable support from the governments in the form of financial and regulatory assistance.

A technology comparable to the solar pump system is the solarised agriculture feeder. The solar feeder is a community-scale solar plant that is connected to the distribution transformers and generates electricity for agricultural needs during certain hours in the morning and for households in the village at night. Such technology can bring the promise of reliable supply of electricity for the rural society as a whole and also reduce the accountability at the farmer's end. Although this is an isolated case that has been taken up by the government of Maharashtra under the Mukhyamantri Saur Krishi Vahini Yojana, in the form of a pilot and whose long-term impacts are yet to be seen, the techno-economic analysis of the project provides a promising picture, as there is no subsidy dependence for farmers implying that they are more cost-effective than solar pumps [19]. It also can sidestep the issues of accessibility to small and marginal farmers that have been observed as an issue with solar pump implementations.

6.2.2 Offshore Wind

India is blessed with a prodigious coastline of 7600 km with fairly shallow waters. This provides optimal sites for offshore wind energy installations. Furthermore, due to the land acquisition issues that wind power plants have had a run in with across many of the southern states, the need for offshore wind plants has been seen as a much-needed alternative [20]. Under the Facilitating Offshore Wind in India (FOWIND) initiative started by the European Union, India's offshore wind energy potential was estimated to be located in eight zones surrounding the states of Gujarat and Tamil Nadu. Offshore wind energy installations at 100 m hub heights could be attributed with a high-capacity utilisation factor of 40–54% [21]. Such studies in conjunction with light detection and ranging studies aimed to instil confidence in the minds of project developers. Given the policy impetus the Offshore Wind Energy policy of

2015 provided, the Ministry of New and Renewable Energy (MNRE) announced a target of 5GW and 30GW installed capacity by 2022 and 2030, respectively. However, the challenges in achieving the targets are largely technological, due to unavailability of historical data and sophisticated weather research and forecasting methodologies to estimate sea breeze patterns that can support precise identification of potential sites. These technological issues are commensurate with the lack of conducive economic conditions due to high upfront capital costs and resultant high cost of evacuation of power. Additionally, the process of implementation is wrought with operational issues on account of cumbersome clearance processes and long gestation period [22, 23].

Thus, without government support in the form of conducive financial assistance, offshore targets cannot be achieved. In keeping with these beliefs, MNRE published an expression of interest for the first offshore wind project of 1 GW in 2018. However, according to a Global Wind Energy Council report published in 2020, the tendering process for the same has been delayed, rendering the targets even more unfeasible [24]. However, as the technology gains in operational and economical credibility, another issue to be wary of is the growing environmental concerns that it poses. Studies across the world have shown that offshore wind projects can affect the surrounding environment in manifold ways. The impacts on the ambient marine life are at their peak during the construction phase [25]. Additionally, the installations can disrupt the ecological habitat at the bottom of the sea and the uppermost layers of the water and increase noise levels [26]. Learning from the European experience and pre-empting environmental dangers to the ecosphere is definitely an important solution to address these environmental concerns.

6.2.3 Wind–Solar Hybrid Parks

In November 2020, the MNRE published a concept note introducing a scheme that foresaw installation of solar panels with wind turbines [27]. Wind–solar hybrid parks are the newest proposition from the MNRE that could allow for more judicious use of land. The National Institute of Wind Energy (NIWE) identified potential sites for wind hybrid parks in the states of Andhra Pradesh, Tamil Nadu, Karnataka, Gujarat, Rajasthan, Madhya Pradesh and Telangana. If these sites were then found conducive for solar park implementations, plans for a hybrid park would be developed. It was announced that only areas with a capacity utilisation factor of more than 30% would be considered for development of parks of size more than 500 MW. The ministry would act as a facilitator or implementing agency, while the park can be developed by a state agency, a private developer, a joint venture or a special purpose vehicle. For the development of every hybrid park, the ministry would be providing central financial assistance to the tune of Rs. 25 lakhs.

The generated power would be evacuated in inter-state arrangements. Power purchase agreements for twenty-five years with arrangement of power procurement and determination of tariff would be designed on the basis of competitive bidding guidelines. These projects would also enjoy the advantage of not being backed down in the event of over injection of electricity into the grid. In case of an unforeseen circumstance in the form of backdown of power, the power generator would be eligible for a minimum generation compensation from the procurer [28]. Only in case of curtailment of power for grid safety, the generator will not be eligible for any compensation. Moreover, there are no additional transmission costs that are required to be incurred as the only cost is that of connecting to the local substation [29].

One major technological advantage of hybrid systems is that it can dilute the previously experienced variability in the grid. This is because the wind and solar generation complements each other in diluting the intermittency of the energy resources [30]. Such a hybrid system would also reduce the need for a storage requirement. Storage is an important demand response strategy that India seeks to make a mainstay subsequently. However, until battery storage can be considered as an economically cost-effective solution, hybrid solutions can help dampen the peak load pressures on electricity demand.

The scheme has identified a whopping 10,800 km^2 of land across various RE rich states for the development of hybrid projects. However, the scheme does not consider the impact of possession of this land or possible impacts the land use change can have on the surrounding environment. Environmental and social impact assessments are necessary for such large-scale projects to be implemented. Furthermore, the efficiency of generation in such large-scale projects is a persistent problem as the amount of land required for installing the infrastructure does not necessarily translate to an equivalent amount of power generation, which exacerbates the problem of land availability and utilisation.

6.2.4 Energy Storage

As the spate of RE capacity addition increases, storage solutions are coming into the spotlight as it balances the power generated and utilised and enables a more flexible dispatch strategy. This is where energy storage can be the solution to unleash the true potential of RE in India. Energy storage implies the storage of previously generated electricity for use at a later time. This technology can be accessed with the use of batteries, pumped hydropower storage and pressurised gas storage. While gas-based storage systems cannot be ascribed as a green solution, India has strongly taken to invest in battery technologies and pumped hydropower storage.

Although there are several battery technologies, the lithium-ion-based battery technology has made major inroads in India. They can help to meet the peak morning and evening demands [31]. In alignment with this, the National Energy Storage Mission was introduced to create a conducive environment for battery storage and fuel efficient RE integration [32]. Initially, applications of battery storage came in the form of retrofitting in existing solar wind hybrid projects in India [33]. Solar Energy Corporation of India (SECI) facilitated a number of proposals for battery storage. One such project is for 1200 MW of solar and wind with guaranteed supply during peak hours that is, the project must supply power for a minimum of six hours during peak times. While the off-peak tariff was determined by SECI at Rs. 2.88/kWh,[1] bids were invited for peak tariff. The bid was won by Greenko at a tariff of Rs. 6.12/kWh for a capacity of 900 MW and ReNew Power at a tariff of Rs. 6.85/kWh for a capacity of 300 MW. While Greenko aims to use pumped hydro as the resource for storage, ReNew Power is to use lithium-ion batteries for a solar and wind hybrid system [34]. This is indicative of the fact that grid-scale storage is underway across the country [31]. An interesting insight that this project brought to the fore was the closing gap between tariff of RE integrated with storage and fossil fuel-based energy which at a recent tender was priced at Rs. 4.24/kWh [35]. Another analysis found that the RE integrated storage price could fall to Rs. 3.4/kWh by 2030 [36].

In keeping with the government's vision of India becoming 'atmanirbhar' or 'self-reliant', plans to set up battery storage were undertaken in 2019. India's expected demand for batteries till 2030 is about 1100 GWh [37]. In accordance with this forecasted demand, NITI Aayog—the government think tank—proposed a plan to seek for proposals across the country for 50 GW capacity of battery plants, incentivising the process with duty waivers and exemptions to battery manufacturers. NITI Aayog would also initiate the tendering process after the identification of locations [38]. The government has taken its well-intentioned initiative closer to realisation through the introduction of a production-linked incentive scheme. Under this scheme, 180 billion USD have been pegged for battery manufacturing which can help in providing round the clock RE in the country. The government would provide a subsidy conditional on the amount of output manufactured from the date of commencement of the project [39]. However, as the government initiatives stoke more investment in battery manufacturing, a greater responsibility falls on ensuring proper disposal of these batteries after the completion of their shelf life. Their disposal in landfills can have long-lasting impacts on the environment [40]. Moreover, battery manufacturing foments the need for intensive mining of rare earth metals that bring with it their own set of socio-environmental and geopolitical issues [41].

The alternative to battery storage is pumped hydropower. The Central Electricity Authority (CEA) published a report in 2017 to announce that India had a pumped hydro potential of 96 GW across sixty-three sites [42]. India currently boasts of more than 5.5 GW capacity of pumped storage hydropower in operation and construction. While this forms a small percentage of the globally installed 140 GW of pumped hydropower storage, this technology has proven credible worldwide [43]. In 2020, Greenko obtained financial assistance to the tune of 495 million USD from foreign private developers for the purpose of developing two integrated projects which would combine 8000 MWh of pumped storage with 2 GW of solar and wind energy each [44]. However, this technology is also not devoid of the environmental and social costs that have plagued hydropower projects for decades. They pose the dangers of deepening issues of displacement of rural communities and worsening seasonal water flows among other issues related to water rights.

---

[1] kWh (kilo watt-hour) is the unit for energy.

6.2.5 Transmission Planning

As RE capacity proliferates, efforts to ensure its integration also have to be increased. Given the temporal variability of RE, it is imperative to improve flexibility of the transmission system. Flexibility alludes to the ability of the grid to adapt and accommodate sudden changes in demand and supply of electricity [45]. Due to the output uncertainty in RE generation, real-time scheduling of power due to inadequate forecasting amenities becomes a challenge. RE also enjoys must-run status, implying that the fossil fuel-based generators serve the net load over and above the base load. Given the fluctuating RE generation and demand, the pressure to ramp up and ramp down fossil fuel-based generators is immense. Quick shut down, ramp up and ramp down is difficult to achieve due to loss in efficiency, equipment deterioration and lifetime reduction. At times surplus generation of RE cannot be counteracted with reduced thermal generation, leading to grid frequency fluctuations. In order to circumvent this issue, the deviation settlement mechanism regulations were introduced by CERC in 2014. These regulations proscribed surplus and under generation of RE beyond certain bands by providing incentives and imposing penalties on the RE generators [46].

Network congestion is the other issue that comes with improper scheduling and forecasting of RE generation. In fact, lack of transmission infrastructure has even led to cancellation of RE projects [47]. The central government has undertaken a number of endeavours to allay these issues. One of the most important of these is the creation of the Green Energy Corridor and Renewable Energy Zones for evacuation of 106.5 GW of RE capacity. The Green Energy Corridor was proposed under the twelfth Five Year Plan period in 2015 to improve intra-state and inter-state transmission of energy from RE-rich areas to the rest of the country. The project was to be implemented by Power Grid Corporation of India (PGCIL) in association with the respective states [48]. The initial phase of the infrastructure development occurred under a regulated tariff mechanism where the cost was to be borne by PGCIL who would also earn a return on equity on these projects [49]. The subsequent phase of the project was opened up to private participation and attracted as much as Rs. 150 billion worth of transmission bids [50, 51]. Until 2019, 2168.20 circuit km was constructed for intra-state transmission, and 2467 circuit km was constructed for inter-state transmission. This upgraded transmission system is capable of supporting a total of 17,757 MVA^2 of substations and aiding in evacuation of 10,261 MW of RE capacity added [52]. Currently, the corridor expansion is still underway in the northern region in addition to the increase in the transmission interconnection capacity between the western and southern regional grids [53]. Interestingly, the Asian Development Bank has pegged the green energy corridor project as unsatisfactory in bearing results of social impact assessment.

---

[2] Mega Volt-Ampere (MVA) is the unit for apparent power that runs in an electrical circuit.

6.2.6 Green Hydrogen

Most recently, India has turned its focus to the unique technology of 'green hydrogen' which has myriad applications. With talks of the very first installation of production facilities in Tamil Nadu, India is turning a new page in its energy transition journey. The technology underscores that if the hydrogen is produced from RE sources or electrolysis, it is termed as 'green hydrogen'. The Prime Minister recently proposed the launching of a National Hydrogen Energy Mission in 2021–22 which would promote the generation of green hydrogen as opposed to the current status quo where most of the hydrogen produced is sourced from fossil fuels in the country [54]. From an economic perspective, an analysis done by TERI highlighted promising prospects of the technology, as hydrogen costs are forecasted to fall by 50% by 2030 [55]. This new technology would also require new institutional frameworks in place. But one cannot throw caution to the wind when interacting with this new technology without pre-empting the challenges the technology may pose. It still has a nascent market and requires establishment of international and national standardisations to ensure guarantees of origin of the hydrogen and other unforeseen issues that are yet to come to the fore [56].

6.3 Economics and the Transition: Energy Darwinism in Play

In 2013, the reports of financial giant Citigroup introduced a term 'Energy Darwinism'. The term alludes to an economic discourse that proposes that as fossil fuel costs keep increasing, the market will naturally gravitate towards becoming greener and cleaner [57]. This implies that RE is naturally becoming more competitive in the market and is translating into becoming investment-friendly and risk-free. Thus, the concept fits RE into the tropes of neoliberal rationality. This conceptual framework, reflective of RE naturally adapting to the market, holds true across the globe in general and in India in particular.

Over the decade that has passed landed cost of electricity from solar photovoltaic (PV) systems has declined by a significant 36% worldwide. The cost of the balance of systems that is the entire PV system can be considered as the main contributor of this decline. IRENA in a report published in 2019 stated that Asia contributed 60% of the new installations in especially China and India [58]. As the solar energy generation is scaling up, India is witnessing an increase in number of cost competitive solar projects. The price discovered through the most recent competitive bidding project was a minimal Rs. 1.99/kWh in a Gujarat auction. The price was quoted by four bidders for 400 MW capacity [59]. The aggressiveness in bidding stood testimony to the fact that the market was trying to keep up with the increasing electricity demand, increasing eagerness of foreign investors, low cost of debt and future expectation of low PV module prices [60]. With SECI acting as the nodal agency and power purchase guaranteed at prices competitive with thermal power, the plan for future capacity addition was cogently drawn out. However, uncertainty looms large as the current government's vision of India becoming self-reliant necessitated imposition of safeguard duty on solar imports until July 2021 and customs duty on solar imports as mentioned in the union budget 2020–21 [61, 62]. This may effectively raise solar tariffs as customs duty as high as 40% are levied on solar modules from April 2022. While the intention of such duties is to provide impetus to domestic manufacturing, it may come at the cost of increasing uncertainty over power offtake as there are growing delays in solar project completions due to signing of power purchase agreements due to possible rise in solar tariffs [63].

The trajectory was similar for wind energy in the country in the last decade. While the turbine quality rating is considerably lower in India in comparison with Denmark, prices have witnessed a drastic fall by 71% between 1984 and 2019 in India [58]. The capacity utilisation factor has also risen from 25% in 2010 to 32% in 2019 alluding to the increase in generation efficiency. However, after the record low tariff of Rs. 2.64/kWh discovered in 2017, the wind sector has fallen into more troubled times. Albeit its presence in the RE market for a longer time, the wind sector has not witnessed the same degree of maturity as the solar sector has. Consequently, in order to allay the concerns of investors and manufacturers, competitive bidding guidelines were amended by the government in 2019 to fasten the process of wind project implementations [64]. SECI is also the nodal agency for implementation of wind power projects. This move has aided in bid quotes to come down to a low range of Rs. 2.4–2.6/kWh as of 2019 [65]. However, these low prices do not necessarily translate to offtake guarantees across the states and have affected small domestic investors who cannot benefit from economies of scale. Persistent concerns related to land acquisition also tend to raise prices on account of the associated risks. Experts state that wind energy projects were more conducive to the feed-in-tariff system wherein State and Central Electricity Regulatory Commissions (CERCs) determined the tariffs for wind projects. In contrast, the reverse auction competitive bidding system introduced in 2017 wreaked havoc with the wind sector. Dissonance over power purchase agreements signed between distribution companies (DISCOMs) and wind generators under the feed-in-tariff system continue even today. The payment dues to generators have been on the rise and have put cash flow pressures on generators and may subsequently affect the credit ratings of these wind projects [66]. Conducive policies that allow for cheaper inter-state transmission could alleviate a lot of the issues the wind sector was facing in 2019. But in the midst of all the existing issues the sector was plagued with, there was a bigger threat looming which reared its ugly head in 2020.

The COVID-19 pandemic that ensued in early 2020 had repercussions across the board. The RE sector also was not spared its wrath. Worldwide it stalled and delayed wind and solar PV installations due to the government mandated lockdowns [67]. However, in India, the tide took a different turn. The sector has shown resilience in face of the virulent adversity as India is expected to double its capacity in 2021 in comparison with the 2020 levels [68]. Notwithstanding this resilience the Indian RE sector has been adversely impacted by the global supply chains [69]. Given the standstill in solar imports from China during the peak of the pandemic, precious time was lost by way of capacity addition. The government mandated lockdowns necessitated countrywide shutdown of all industries. Construction of solar projects could only resume from April 2020, but the lack of labour, logistical challenges and delays in site inspection made progress tougher [70]. On the flipside, a dip in industrial production translated to a 40% decline in energy demand within the nation. A ripple effect followed as the mismatch in electricity demand and electricity scheduled by DISCOMs pushed them into financial doldrums as they stood to lose revenue worth 12 billion USD. As a means to deal with the unprecedented situation, electricity supply had to be curtailed, and this placed intermittent RE at a disadvantage. Given the financial constraints of DISCOMs, the payment dues to RE generators have also been burgeoning [69, 71]. This in turn has accreted risk and uncertainty with regard to investments in the sector. The union budget of 2020–21 also recognised DISCOM viability as an essential facet to focus on and allocated a Rs. 3.05 trillion package to financially disturbed DISCOMs. The budget also reinstated faith in solar by allocating Rs. 2500 crore to the SECI and Indian Renewable Energy Development Agency [62].

In addition to the troubles brought on by the pandemic, the sector has been witness to an ongoing discussion on integration of balancing costs for RE. This implies that RE should not be deemed as having zero marginal cost but must be held accountable for maintaining the health of the grid. Such integration would optimise scheduling and dispatch decisions made by the DISCOMs on the basis of accurate forecasting. It would appropriate values to variability and curtailment of RE, based on flexibility of thermal generation and other sources of power generation and peaks and off-peaks in load curves. It would use high resolution geographic maps to ascertain RE potential. Advanced models would also take into consideration historical data to understand technical, operational and market constraints [72]. Thus, integration of system costs requires investment on grid infrastructure systems that can condone advanced modelling and bolster grid flexibility. In the long run, if infrastructure can be improved and demand response modelling can be practiced ubiquitously beyond academic and grey literature, we can successfully ensure least cost analysis for DISCOMs [73]. As a frontrunner, the government of Andhra Pradesh has taken a stand to bear balancing costs associated with 10GW addition of solar capacity for thirty years. This would assuage the DISCOMs' financial conditions. The costs are aimed to be reduced by development of a forecasting model [74]. Thus, accountability towards these costs can be introduced by sharing costs among all stakeholders. The forecasting and scheduling regulations aim to provide a framework to initiate this process.

De [75] says that while RE capacity addition will pick up in the long run, the trajectory will not be devoid of hurdles. This will be essentially due to investor sentiment given the risks and costs associated with RE post pandemic. Thus, while the concept of 'Energy Darwinism' may have boded a promising beginning for a competitive RE market, support for RE investment must not only be driven by market forces but other sources too. As the union budget of 2020–21 clearly elucidates, given the '*Atmanirbhar Bharat*' stance of the government, the need for economic stimulus is now more than ever. A recent standing committee report by MNRE acknowledges that in order for India to be successful in achieving its 175 GW target by 2022, it still needs to invest Rs. 2.6 trillion on RE capacity addition which is a long way from the Rs. 823 billion investment the country has been doing on an average over the last five years [76]. The path to economic recovery for India clearly moves along the same trajectory as green energy investment [77].

6.4 Regulatory Frameworks

Strong regulatory frameworks form the bedrock of RE traction in the country. They are one of the main cornerstones of the Indian RE sector that have not only sustained but also ensured the strengthening of the sector, making India a stalwart in the energy transitions story. There are four regulatory frameworks that can imminently or already have proven to bring key watershed moments in the sector.

6.4.1 Renewable Purchase Obligation Framework

The Renewable Purchase Obligation (RPO) framework was mandated under the Electricity Act, 2003, Section 86(1)(e) and the National Tariff Policy, 2006, Section 6.4 [78]. The State Electricity Regulatory Commissions (SERCs) were mandated to determine a minimum percentage of total purchase by large energy consumers like DISCOMs, open access consumers and captive power producers,[3,4] that would be sourced from RE. These solar and non-solar percentages were ascertained as RPO targets.

Figure 6.4 provides a list of RPO targets determined by the Ministry of Power and three of the RE-rich states in the country. The state nodal agencies, namely the state energy development agencies, were delegated the responsibility of holding the obligated entities accountable by monitoring their yearly RPO compliance. The SERCs hold the power to penalise obligated entities in the case of incomplete yearly RPO compliance.

In conjunction with the RPO framework, the Renewable Energy Certificates (RECs) were introduced in 2010 to allow for accounting for RPO compliance without actual transfer of green energy. RE generators under the purview of this mechanism would generate green energy and sell the energy at the same tariff as the electricity from fossil fuels, while rest of the cost of green energy was to be obtained from the sale of RECs issued for every 1 MWh[5] energy generated. The price of the RECs

---

[3] Captive power producers are privately owned power generating plants that consume a minimum of 51% of the energy produced and sell the rest.

[4] Those who need to abide by the RPO targets are known as obligated entities.

[5] Mega Watt-hour.

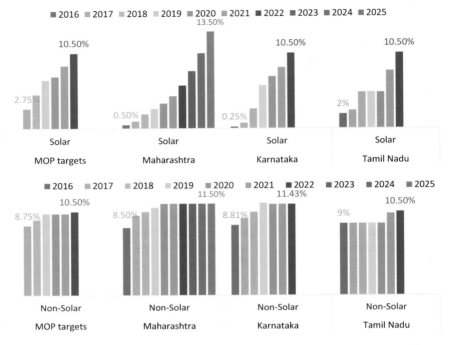

Fig. 6.4 State-wise solar and non-solar targets. *Source* Data collated by author from Prayas (Energy Group) Renewable Energy data portal [79] and Tamil Nadu RPO regulations (In this figure, each year denotes a financial year, that is 2016 implies 2015–16 or the year starting from 1 April 2015 to 31 March 2016)

would be determined in the power exchanges and would be susceptible to demand and supply of the same. This would circumvent the issue of states with lower RE potential finding it difficult to achieve their RPO targets. The RPO and REC frameworks provided the sector much-needed boost in the early phase of the previous decade when RE tariffs were not competitive with fossil fuel-based energy tariffs.

As RE achieved grid parity in 2017, a stream of discussion emerged on the question of merging solar with non-solar RPOs. The Maharashtra state-owned DISCOM filed a petition at the Maharashtra Electricity Regulatory Commission (MERC) with this as their prayer. As all sources of RE became competitive price-wise, the onus could be on the obligated entities to decide which RE source they wished to fulfil their RPO targets with [71]. However, the Commission found that merging RPO targets would send an incorrect signal to the market and dismissed the petition. But this school of thought in fact can ensure more consistent RPO compliance by obligated entities and should in the long-term gain traction in the policy circles.

As transmission planning is underway and obligated entities' access to RE is increasingly eased, the other growing concern has been the obsolescence of the REC mechanism. As Fig. 6.5 illustrates, demand for RECs has consistently been lower than the supply in the last decade. This consistent gap highlights the constant pressure

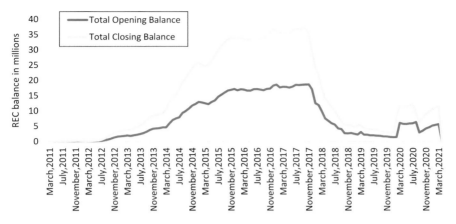

Fig. 6.5 Monthly renewable energy certificate opening and closing balance from 2011 to 2021. *Source* Data collated by author from REC Registry Website

RE generators under the REC mechanism have been in, as the fear of expiration of unsold RECs gets realised. The green day ahead and green term ahead markets that were introduced in the power exchange market may further reduce the requirement of RECs as more obligated entities opt to purchase green energy directly from these markets [80]. A clear indicator of the RECs getting outmoded is the most recent order by CERC, in which the minimum (floor) price at which RECs (both solar and non-solar) can be bought at is zero and the maximum (forbearance) is Rs. 1000/MWh [81]. With such low prices for selling RECs, RE generators could be dissuaded from coming under the purview of the mechanism. This is indicative of the central autonomous institution taking stock of the current market trends, clearly alluding to the gradual phasing out of the REC mechanism. Moreover, the same floor and forbearance price of solar and non-solar RECs is the first step towards merging RPO targets in the long run.

RPO targets have received resounding precedence across the country as all states publish their respective RPO regulations periodically. After the central government announced the 175 GW target by 2022, the focus on stringent RPO compliance was further emphasised. For instance, in order to promote a consistent cycle of accountability, the central government energy minister in 2019 stated that all states must inform the central government of their RPO compliance on a quarterly basis [82]. The pandemic, however, brought on a speed bump in 2020, as RPO compliance rules were relaxed in various states so as to provide obligated entities some relief. For instance, in Karnataka, time for RPO compliance was extended until end of December 2020, and in Punjab, a reduction in the RPO targets was suggested [83, 84].

Currently, the MNRE's National portal for RPO Website provides an overview of the way forward for the successful achievement of RPO targets through a centralised monitoring mechanism that can ensure transparency in the compliance process [85]. Some individual states also have RPO portals to ease the tracking of the compliance process. All these modes of tracking clearly alludes to a digitised future for the framework which would require cooperation among state nodal agencies and the ministry.

6.4.2 Renewable Energy-Based Open Access

The Electricity Act, 2003 introduced the phenomenon of open access (OA) with modalities as depicted in Fig. 6.6. But the framework was not implemented in letter and spirit [86]. While RE-based OA has immense potential in the country, it did not see a lift off for quite a lot of time for a number of reasons. First, many individual states imposed various regulatory hurdles to deter OA in order to protect DISCOMs. This is because, all OA consumers are large consumers who consume more than 1 MW of power. If a large industrial or commercial consumer opts for OA, it would move out of the state DISCOMs' purview. This would result in the utility to lose out on a substantial quantum of revenue from electricity tariff while being left with the low paying domestic and subsidised rural and agricultural consumers. Thus, the state governments' response may reflect the belief that migration of such high paying consumers in the name of choice and competition would have adverse effects on the DISCOMs' finances. Second, the considerably high OA charges that an interested consumer had to pay also acted as a deterrent. Finally, the operational issues related to obtaining approval for these processes are arduous and often end in failure as DISCOMs must provide no objection certificates [87, 88]. As a testimony to this, in

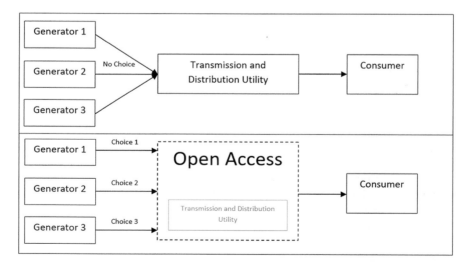

Fig. 6.6 Modalities of open access. *Source* Illustration by author, based on the Electricity Act, 2003 [90]

2018, the solar OA market was pegged at a meagre 2894 MW which is just 12% of the total solar capacity in the country [89].

In order to avail the option of OA, a consumer has to pay a number of fees and charges.[6] But since Section 86(1)(e) of the Electricity Act, 2003 mandates promotion of electricity generation from RE source, various states introduced promotional concessions to aid in the pickup of electricity generation from RE [90]. Concessions are offered in the form of waivers and exemptions in various OA charges like the cross-subsidy surcharge and transmission and wheeling charges. These concessions initially enabled RE-based OA transactions to gain traction in many RE-rich states like Maharashtra, Karnataka and Tamil Nadu. But the mechanism has reached a point of inflection in the last two to three years, as more states are opting to remove these concessions. As RE grows more competitive, there is a need for smoother transitions in OA charges with gradual phasing out of these concessions [91]. The concessions are increased and reduced in an inconsistent manner, not reflective of any trends in the different states which make it difficult to plan for investment in long-term OA projects.

Moreover, heavy regulation of third-party sales and captive power producers outside the purview of DISCOMs has stunted the growth of RE. There lies a conflict of interest in regulation, especially if approvals are provided by DISCOMs. As an instance, substantial number of regulatory cases were filed in the Maharashtra Electricity Regulatory Commission by OA consumers and RE generators due to denial of OA by DISCOMs [71]. A way forward can be for DISCOMs to look at sales migration not as tariff revenue lost but as a source of revenue for services provided for the use of wires for wheeling energy [92].

One of the specific features of the OA regulation that requires special attention is the banking mechanism. This policy mechanism was introduced by regulatory commissions to assuage the tussle between must-run status that RE generators enjoy and the RE curtailments on account of surplus energy production during solar and wind peaks. Banking simply implies that in the event of surplus energy produced that the OA consumer does not need at that instance, it can be banked with the DISCOM for use at a later date, usually within a year. The consumers can use this service with the levy of a banking charge. However, this charge is levied in an ad hoc way devoid of any specific principles. Prayas (Energy Group), a nonprofit think tank, proposed a mechanism that can allow accurate banking charges that are reflective of real-time energy prices [86, 91]. Thus, for RE-based open access to gain traction, a level playing field for the open access consumers and the DISCOMs is a necessity.

6.4.3 Net Metering Regulations

The National Solar Mission posed a target of 20 GW of grid connected solar capacity by 2022. Under this mission, a scheme was launched namely, Rooftop PV and Small

---

[6] For details of these charges and the various modalities associated with them, see [86].

Scale Solar Generation Programme. This scheme did not receive much traction in the initial phase of the National Solar Mission between 2010–2013. Rooftop PV capacity was close to non-existent during that period on account of lack of clarity on applicable charges, guidelines on energy accounting and the technical standards required [93]. To incentivise the small-scale sector, the net metering mechanism was drafted. This mechanism in fact is a less stringent kind of banking service that is allowed for rooftop solar PV installations. This framework allows the solar system owner to only pay for the 'net' energy consumed. If the excess energy fed into grid within the billing period is more than the energy consumed from the grid, the excess energy is carried forward to the next billing period. A bi-directional meter connected to the system keeps track of the net energy exchange, thereby condoning calculation of 'net' energy consumed. This framework promises extensive savings on energy bills for residential consumers. Moreover, the framework circumvents the use of batteries as it essentially allows the consumer to 'bank' the excess energy with the DISCOM for use on a later date, within the billing period. For this mechanism to be successful, there was need for a robust regulatory and commercial framework to be introduced nationwide.

As the concept evolved at the central level, around twenty-one states issued their net metering regulations that would allow low end residential consumers access to this framework [94]. By 2017, all states and union territories had come out with their net metering regulations [95]. The regulations varied in the capacity allowed for installation, the tariff to be paid by DISCOMs for excess energy sale and the amount of excess energy to be condoned for billing and the billing period, among other specifics [94].

Notwithstanding the growing policy support for the framework, as Fig. 6.7 highlights, the uptake of the rooftop solar capacity was sluggish. The reasons for this slow growth ranged from operational and implementation issues that tended to raise transactional costs. Additionally, similar to the case of open access, DISCOMs took to the framework antagonistically, on account of fear of loss of tariff revenue. A relook at the framework was required to allow a balanced stance towards all stakeholders [96].

In alignment with this issue, the government notified the Electricity (Rights of Consumers) Rules 2020, in which net metering was allowed for those willing to install rooftop solar systems up to 10 kW while gross metering would be applicable for all loads upwards of 10 kW [97]. Gross metering enables payment to consumer for energy generated and fed into the grid at a feed-in-tariff, while the consumer continues to pay the regular retail tariff for their energy consumption to the DISCOM. This change was mandated, keeping in mind the dismal financial health of DISCOMs. The key question that this new mandate raised was whether the feed-in-tariff will be lesser than the retail tariff or will it be subject to change over the years [98, 99]. This would deal a blow to the solar developer industry as small- and medium-scale enterprises and larger industrial consumers with load more than 10 kW looking to cash in on the benefits of net metering may end up being outside its purview. The rooftop solar sector had only started picking up on account of net metering, and such a mandate could threaten to thwart that growth. Given the vehement opposition this

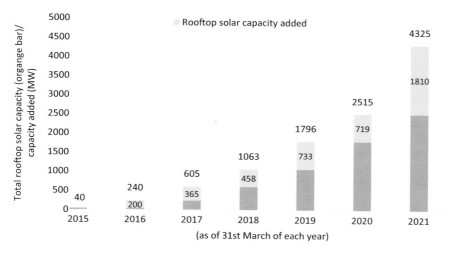

Fig. 6.7 Rooftop solar capacity from 2015–2021. *Source* Data collated by author from various reports and websites [1, 95] and [79] (In this figure, each year denotes a financial year, that is 2016 implies 2015–16 or the year starting from 1 April 2015 to 31 March 2016)

new rule has faced, the Ministry of Power has been forced to reconsider it, and the rule is back on the drawing board [100]. The larger concern is the effect this back and forth on policy will have on the achievement of the 40 GW rooftop solar target by 2022 of which a little more than 4GW has been achieved. As time ticks away, achieving the target in less than one year starts to look more and more prodigious unless rapid action is taken soon.

6.4.4 Forecasting and Scheduling Regulations

As the tussle between the must-run status of RE generators and curtailment of the RE generation for grid stability heightens, the importance of forecasting and scheduling regulations also increases [101]. As RE penetration increases, there is a need to acknowledge how much of it is being fed into the grid and how much can be condoned for a healthy grid. The Deviation Settlement Mechanism (DSM) provides a carrot and stick approach to generators by linking monetary penalties/incentives with the grid frequency, as over/under injection of electricity affects the grid frequency.

As a farsighted strategy, the CERC issued the forecasting and scheduling regulations for RE projects connected to the inter-state transmission system in 2015. The procedure for the same was finalised in 2017, followed by which almost all states came out with their respective forecasting and scheduling regulations [91]. However, the regulations have experienced considerable teething problems across the country. In Maharashtra for instance, after the regulations were notified in 2018, nearly 50 regulatory petitions were filed at the SERC in a matter of two years [71]. These petitions raised issues related to the lack of clarity in modalities related to the DSM, concerns over the must-run status of RE generators and additional DSM charges.

By experience, a disadvantage for generators has been the limitation on the number of revisions in the schedule they are allowed. Currently, regulations allow a total of sixteen revisions in twenty-four hours, once in every 1.5 h. However, during periods of high variability, these many revisions may not be enough for generators who would then have to bear the cost of deviation by way of DSM charges. Thus, increasing the number of revisions in the schedule insofar as it does not affect grid safety can provide generators some respite [101].

Given the limited accuracy in forecasting and the variability of RE generation, the state regulations have allowed an error band of $(\pm)10$ or $(\pm)15\%$ in generation from the forecast to not be penalised with DSM charges. This provides an implicit opportunity to RE generators to take corrective measures so as to ensure deviation does not exceed this band. But the cost of the deviation up to $(\pm)10$ or $(\pm)15\%$ is borne by the DISCOMs in most states [91]. An approach which makes the generators bear this cost can also provide them the incentive to reduce the error in deviation, therefore helping in balancing out the costs. Maharashtra has adopted such a practice with positive reviews currently.

Thus, there is a need for regulatory commissions across the states to review and amend the regulations and incorporate the operational experience to get rid of the teething problems. Additionally, they must also tighten the error bands and hold the generators accountable for the costs of deviation. Not doing so, would invariably raise balancing costs of RE. From an infrastructural perspective, the way forward can be rendered wherein as forecasting technology improves and achieves precision these deviation costs would also begin to decline.

6.5 Just Energy Transitions: The Way Forward

India has stood as an exemplar as it has consistently been on the frontline, helming the energy transitions story. But the elephant in the room is the social and environmental costs that must be accrued for RE implementation, and yet is swept under the rug. Energy transitions can perpetuate a pre-existing set of winners and losers [102]. Winners can be those who enjoy the benefits of clean energy, reduction in emissions by fossil fuel, and employment opportunities due to innovations in the energy system. Losers can be those who would lack access to these opportunities or are adversely affected by the implementation processes. In contrast, a just energy transitions framework for RE infrastructure can be conceptualised as not only comprising socio-technical interventions but also as being sociopolitical in nature [103]. A failure to acknowledge this sociopolitical character of RE technology can lead to unjust energy transitions on account of uninformed policymaking. Hence, there is a need to add 'just' in the clean energy transitions discourse in India.

The movement for just transitions was steered by the labour unions hailing from low-income coloured communities and has been explained as *'a principle, a process and a practice*' [104]. This alludes to the coproduction of social relations in association with the energy systems transition, thereby deeming such transitions as just or unjust. In fact, one can extrapolate these social relations to allude to the coproduction of the ecosphere in association with RE technologies. Thus, energy transitions must be analysed through a more holistic lens that acknowledges the sociological and the overarching ecological hazards RE poses. This implies that we consider RE in connection with the political economy of ecological movements [105], thereby balancing development goals with environmental justice issues.

Recent scholarship on energy across different countries suggests that RE can reproduce social and ecological inequalities. In Southern Catalonia in Spain, there exists a vexed relationship between energy and agrarian lives to the extent that they co-produce each other, while social inequalities remain unchanged, notwithstanding energy transition from nuclear to wind [106]. In another instance, one can witness how the use of solar panels and installation of high voltage transmission wires in a Peru village adversely affected the social fabric of the village [107]. The limited presence of the new technology produced considerable degree of polarisation. In the Indian context, the just energy transition has made inroads only in the form of acknowledgement of the phasing out of coal and the repercussions this can have on the millions of coal workers and the possible reproduction of unequal socioeconomic structures [108, 109]. But the larger concern must be to question why India while citing ambitious RE targets has also continued to push for extraction of coal instead of phasing it out [110]. Additionally, we should not be reproducing the injustices meted out under fossil fuel economies with large-scale RE implementations [111]. While RE propitiates many, it also has been shot down by sceptics who believe that it may not always warrant relief from the social inequities and in fact may perpetuate an unjust energy transition. Thus, RE may not be deemed as the silver bullet after all if it ends up reinforcing pre-existing social inequalities. In India, land grabbing or deeming lands as 'backwards' to provide them for solar park development is a common feature as is accounted in states like Gujarat and Andhra Pradesh [112]. This clearly alludes to the effects rampant RE construction and development have on communities that are in the vicinity of these projects who are rendered dispossessed. Another instance is the contestation of local communities to RE projects like the wind project in the Bhimashankar Wildlife Sanctuary due to the threats the project posed to their agrarian practises and the ecology of the landscape [113].

The overarching point of the just energy transitions movement is to restructure the energy transitions story while taking cognisance of the existent power relations that have allowed the current status quo of unequal structures to exist. Such analyses should aim to make these power relations a status quo ante. This is imperative for medium- to long-term planning for RE infrastructure in the country. As large-scale solar parks become more common, the operational experience has thrown light on extremely dire impact of the implementations on the communities and land in the region. Pavagada solar park in Karnataka is a classic example that depicts the power relations that exist and that may impact future RE development. While the Karnataka government celebrated the success story of the largest solar park in the country, ethnographic studies showed the dissonance that existed at grass roots level and lay bare the power relations in play. Having been constructed over 13,000 acres of land, the park left farmers unhappy with the rent they were provided, concerned over the imminent decline in fertility of soil after the twenty-eight years of the park life and irked those who claimed coercion was involved in lease arrangements [114].

As land acquisition became a bottleneck for solar project developers, a solution was sought in floating solar projects. This is another instance that puts RE in direct conflict with the environment. As of 2019, more than 1.7 GW capacity of floatovoltaic projects were under construction, and 2.7 MW were operational. However, a study carried out by the think tank TERI highlighted the stress these projects can impose on the freshwater ecosystem, triggering chain reactions and adversely affecting aquatic animals. As solar panels floating on water bodies reduce the extent of sunlight entering the ecosystem, they can disrupt carbon cycles or worse lead to deep-water anoxia that is absence of oxygen [115].

All these studies beg the question, if RE indeed wishes to be the silver bullet, infrastructure in the form of policies and regulations must be put in place that mandate social and environmental impact assessments for RE projects. While Telangana mandated the need for environmental clearance for setting up solar plants, a legally binding notification is the need of the hour [116, 117]. However, the draft Environmental Impact Assessment 2020 published by the Ministry of Environment, Forests and Climate Change excludes RE sources from under its purview. Life cycle impacts of these technologies must be a crucial facet of these assessments, as post-decommissioning of solar plants, discarded panels which contain dangerous chemicals like cadmium, chromium and lead can leach the soil. There is a need for assessments that conduct socio-ecological studies of the region without condoning exemption from environmental clearances, identifying and demarcating no-go zones, and conducting public consultations. Additionally, carrying out sectoral analysis in addition to case-to-case-based analysis is imperative.

Currently, the National Green Tribunal is responsible for handling and redressing environmental justice-related issues in the country. Social and environmental impact issues related to RE projects also fall under its purview of environmental justice. However, the Tribunal has been claimed to not have been very effective in carrying out its powers [118]. Streamlining the process of ensuring just energy transitions by first and foremost constructing regulatory frameworks is the need of the hour to provide RE development economic viability and certainty in the long run. A larger question that is imminent is whether large-scale RE is indeed the requirement or can the country curb its energy demand and seek infrastructural focus on a more decentralised form of electricity system sourced from clean energy.

India is at a crucial crossroads as it accelerates its energy transition story. The key cog in the wheel for a smooth transition is the infrastructure—that is currently in place and that which can be portended in the near future. The Indian government with its various policies and schemes in the last five to six years has made its stance on RE clear—it is here to stay. While the sector has displayed a potentiality in achieving the country's sustainable development goals seven and thirteen, it has also been a witness to technological, economic, regulatory and environmental constraints. Research and development to introduce new green technologies as alternatives to fossil fuel-based technologies must be considered in tandem with pushing for economies of scale that can make these technologies competitive; implementation of regulatory frameworks that make the traction of these technologies easier; and taking into consideration the impact these technologies can have on the environment (human and extra-human nature). Infrastructural developments in RE cannot be undertaken in silo without taking into cognisance the larger Indian political economy that balances development goals along with the dangers it poses to the ecosphere. In the long run, RE can be deemed a panacea for national development and the climate crisis only if it is acknowledged as a socio-technical intervention and not a technical intervention alone. Such treatment can pave the way for establishment of meaningful RE infrastructure and its subsequent effective management.

References

1. MoP. (2021, February 28). *Ministry of Power*. Retrieved March 25, 2021, from https://powermin.gov.in/en/content/power-sector-glance-all-india
2. MNRE. (2021, February 28). *Physical Progress*. Retrieved March 25, 2021, from https://mnre.gov.in/the-ministry/physical-progress
3. Safi, M. (2016, December 22). India plans nearly 60% of electricity capacity from non-fossil fuels by 2027. *The Guardian*. Retrieved from https://www.theguardian.com/world/2016/dec/21/india-renewable-energy-paris-climate-summit-target
4. PTI. (2020, January 31). India to have 450 GW renewable energy by 2030: President. *The Economics Times*. Retrieved from https://economictimes.indiatimes.com/small-biz/productline/power-generation/india-to-have-450-gw-renewable-energy-by-2030-president/articleshow/73804463.cms
5. Roblin, S. (2016). Solar-powered irrigation: A solution to water management in agriculture? *Renewable Energy Focus, 17*(5), 205–206.
6. Ramos, J., & Ramos, H. M. (2009). Solar powered pumps to supply water for rural or isolated zones: A case study. *Energy for Sustainable Development, 13*(3), 151–158.
7. García, A. M., Gallagher, J., McNabola, A., Poyato, E. C., Barrios, P. M., & Díaz, J. A. (2019). Comparing the environmental and economic impacts of on- or off-grid solar photovoltaics with traditional energy sources for rural irrigation systems. *Renewable Energy, 140*, 895–904.
8. Ministry of Agriculture & Farmers Welfare. (2019, February 12). *KUSUM Scheme*. Retrieved from Press Information Bureau: https://pib.gov.in/Pressreleaseshare.aspx?PRID=1564057
9. Dharmadhikary, S., Bhalerao, R., Dabadge, A., & N, S. (2018, September). *Understanding the Electricity, Water & Agriculture Linkages Understanding the Electricity, Water, Agriculture Linkages Volume 1: Overview*. Retrieved from Prayas (Energy Group): https://www.prayaspune.org/peg/publications/item/395.html
10. Das, N., Dabadge, A., Chirayil, M., Mandal, M., & Josey, A. (2019). *Elephant in the Room: Implications of subsidy practices on DISCOM finances*. Prayas (Energy Group).
11. Krar, P. (2016, August 17). Haryana to offer 90% subsidy to promote solar water pumps. *The Economic Times*. Retrieved from https://economictimes.indiatimes.com/news/politics-and-nation/haryana-to-offer-90-subsidy-to-promote-solar-water-pumps/articleshow/53739936.cms?from=mdr
12. Prasad, N. T. (2020a, July 23). Andhra Pradesh Amends 10 GW Agricultural Solar Program to Mitigate Cash Flow Issues. *Mercom India*.
13. Chandra, K. K. (2018, April 14). Two years after it was launched, the world's first solar cooperative has transformed Gujarat's Dhundi village. *The Hindu*. Retrieved from https://

www.thehindu.com/society/two-years-after-it-was-launched-the-worlds-first-solar-cooperative-has-transformed-gujarats-dhundi-village/article23528444.ece
14. Gupta, E. (2019). The impact of solar water pumps on energy-water-food nexus: Evidence from Rajasthan, India. *Energy Policy, 129*, 598–609.
15. Shah, T., & Chowdhury, S. D. (2017, June). Farm power policies and groundwater markets: Contrasting Gujarat with West Bengal (1990–2015). *Economic & Political Weekly, 52*(25–26).
16. Sahasranaman, M., Kumar, D., Bassi, N., Singh, M., & Ganguly, A. (2018). Solar irrigation cooperatives: Creating the Frankenstein's monster for India's groundwater. *Economic & Political Weekly, 53*(21), 65–68.
17. Kumar, M. D., Raman, S. K., Singh, M., Bassi, N., Sivamohan, M., & Kumar, S. (2020, October 26). The Solar Fiasco: How the sun will eclipse the 'solar advantage'. *LinkedIn*. Retrieved from https://www.linkedin.com/pulse/solar-fiasco-how-sun-eclipse-advantage-dinesh-kumar/?trackingId=YKsjNKH6TeKTOqWGAtnpgw%3D%3D
18. Sahasranaman, M., Kumar, M. D., Verma, M. S., Perry, C. J., Bassi, N., & Sivamohan, M. V. (2021, March 13). Managing groundwater–energy Nexus in India: What will solar pumps achieve? *Economic & Political Weekly, 56*(11).
19. Gambhir, A., & Dixit, S. (2018, December 20). *Powering Agriculture Via Solar Feeders*. Retrieved November 23, 2020, from Prayas Energy Group: https://prayaspune.org/peg/publications/item/400-powering-agriculture-via-solar-feeders.html
20. Dash, P. K. (2019). *Akshay Urja: Offshore Wind Energy In India*. Delhi: Ministry of New and Renewable Energy. Retrieved from https://mnre.gov.in/img/documents/uploads/2e423892727a456e93a684f38d8622f7.pdf
21. Kumar, R., Stallard, T., & Stansby, P. K. (2021). Large-scale offshore wind energy installation in northwest India: Assessment of wind resource using Weather Research and Forecasting and levelized cost of energy. *Wind Energy, 24*, 174–192.
22. Jain, S. (2020, May 11). Why India needs offshore wind to blow coal out of the water. *REcharge*. Retrieved from https://www.rechargenews.com/transition/why-india-needs-offshore-wind-to-blow-coal-out-of-the-water/2-1-806057
23. Gulia, J., & Jain, S. (2019). *Offshore Wiind Energy in India: A territory ready to be explored*. Delhi: JMK Research Analytics. Retrieved from https://jmkresearch.com/wp-content/uploads/2019/10/Offshore-Wind-Energy-in-India_JMK-Research.pdf
24. Mercom India. (2020, August 12). India's Offshore Wind Target Unfeasible, Says GWEC Report. *Mercom India*. Retrieved from https://mercomindia.com/india-offshore-wind-target-unfeasible/
25. Nazir, M. S., Ali, N., Bilal, M., & Iqbal, H. M. (2020). Potential environmental impacts of wind energy development: A global perspective. *Current Opinion in Environmental Science & Health, 13*, 85–90.
26. Kumar, J. C., Kumar, D. V., Baskar, D., Arunsi, B. M., Jenova, R., & Majid, M. (2020). Offshore wind energy status, challenges, opportunities, environmental impacts, occupational health, and safety management in India. *Energy & Environment*.
27. MNRE. (2020). *Concept note for development of wind park/wind solar hybrid park*. Delhi: Ministry of New and Renewable Energy. Retrieved from https://mnre.gov.in/img/documents/uploads/file_f-1605265655087.PDF
28. Prasad, N. T. (2020b, October 22). Solar-wind hybrid projects will stand cancelled if SECI fails to sign PSA within six months. *Mercom India*. Retrieved from https://mercomindia.com/solar-wind-hybrid-projects-stand-canceled/
29. Das, A., Jani, H. K., Nagababu, G., & Kachhwaha, S. S. (2020). A comprehensive review of wind–solar hybrid energy policies in India: Barriers and recommendations. *Renewable Energy Focus*.
30. Solomon, A. A., Kammen, D. M., & Callaway, D. (2016). Investigating the impact of wind–solar complementarities on energy storage requirement and the corresponding supply reliability criteria. *Applied Energy, 168*, 130–145.
31. Deorah, S. M., Abhyankar, N., Arora, S., Gambhir, A., & Phadke, A. (2020). *Estimating the Cost of Grid-Scale Lithium-Ion Battery Storage in India*. Lawrence Berkeley National Laboratory.

32. MNRE. (2018, August 9). *National Energy Storage Mission*. Retrieved March 12, 2021, from Press release: https://pib.gov.in/newsite/PrintRelease.aspx?relid=181698
33. Kenning, T. (2018, September 19). India's first solar-wind hybrid to be retrofitted with storage after year of strong wind. *PV Tech*.
34. SECI. (2020, February 28). *Results of e-bidding and e-RA for 1200 MW ISTS-connected RE Projects with Assured Peak Power Supply (ISTS-VII)*. Retrieved March 12, 2021, from Solar Energy Corporation of India: https://seci.co.in/show_whats_new.php?id=1021
35. Dutta, S. (2020, February 10). Solar storage tariff spells trouble for coal. *The Times of India*.
36. ETEnergyWorld. (2021a, January 22). Battery storage cheaper than new coal power plants: Analysis. *ETEnergyWorld*.
37. Hans, A. (2020, December 7). *Making India 'Atma-Nirbhar' in advance battery storage*. Retrieved March 13, 2021, from http://niti.gov.in/making-india-atma-nirbhar-advance-battery-storage
38. Singh, S. C. (2019, June 11). Companies may soon be invited to set up battery plants. *The Economic Times*.
39. Swain, D. L., & Nair, E. (2021, January 5). Analyzing The PLI scheme for battery manufacturing. *Mondaq: Connecting Knowledge and People*.
40. Kala, S., & Mishra, A. (2021). Battery recycling opportunity and challenges in India. *Materials today: Proceedings*. https://doi.org/10.1016/j.matpr.2021.01.927
41. Sakunai, T., Ito, L., & Tokai, A. (2021). Environmental impact assessment on production and material supply stages of lithium-ion batteries with increasing demands for electric vehicles. *Journal of Material Cycles and Waste Management, 23*, 470–479.
42. CEA. (2017). *Minutes of the meeting-pump storage plants*. New Delhi: Central Electricity Authority.
43. Buckley, T., & Shah, K. (2019). *Pumped Hydro Storage in India: Getting the Right Plans in Place to Achieve a Lower Cost, Low Carbon Electricity Market*. Institute for Energy Economics and Financial Analysis.
44. The International Journal on Hydropower and Dams. (2020, May 14). Integrated pumped-storage schemes for India. *The International Journal on Hydropower and Dams*. Retrieved from https://www.hydropower-dams.com/news/integrated-pumped-storage-schemes-for-india/
45. Das, P., Mathuria, P., Bhakar, R., Mathur, J., Kanudia, A., & Singh, A. (2020). Flexibility requirement for large-scale renewable energy integration in Indian power system: Technology, policy and modeling options. *Energy Strategy Reviews, 29*.
46. CERC. (2014). *Central Electricity Regulatory Commission (Deviation Settlement Mechanism and related matters) Regulations, 2014*. Central Electricity Regulatory Commission.
47. Bridge to India. (2020). *Transmission Planning, Implementation and Status*. Bridge to India.
48. MoP. (2015, March 12). *Green Energy Corridor Project*. Retrieved March 11, 2021, from Press Information Bureau: https://pib.gov.in/newsite/PrintRelease.aspx?relid=116890
49. CERC, Petition No. 172/TT/2018 (Central Electricity Regulatory Commission August 6, 2019). Retrieved from http://www.cercind.gov.in/2019/orders/172-TT-2018.pdf
50. PGCIL. (2021, March 11). *Tariff-Based Bidding to Award Green Corridor for Solar Parks*. Retrieved March 12, 2021, from https://www.powergridindia.com/tariff-based-bidding-award-green-corridor-solar-parks
51. Joshi, A. (2020, May 30). We are seeing slower revenue, logistics issues: Manish Agarwal, CEO, Sterlite Power Solutions. *ET Engergyworld*.
52. Prateek, S. (2019b, July 12). Nearly 10 GW of renewable capacity added to green energy corridor: RK Singh. *MERCOM India*.
53. Asian Development Bank. (2020, December 3). *India: Green Energy Corridor and Grid Strengthening Project*. Retrieved March 12, 2021, from https://www.adb.org/projects/44426-016/main#project-pds
54. Frangoul, A. (2021, March 10). India is turning to 'green hydrogen' in a bid to decarbonize its economy. *CNBC*. Retrieved from https://www.cnbc.com/2021/03/10/india-turns-to-green-hydrogen-in-a-bid-to-decarbonize-its-economy.html

55. Hall, W., Spencer, T., Renjith, G., & Dayal, S. (2020). *The potential role of hydrogen in India: A pathway for scaling-up low carbon hydrogen across the economy*. Delhi: The Energy Resources Institute. Retrieved from https://www.teriin.org/sites/default/files/2020-12/Report%20on%20The%20Potential%20Role%20of%20Hydrogen%20in%20India%20%E2%80%93%20%27Harnessing%20the%20Hype%27.pdf
56. Abad, A. V., & Dodds, P. E. (2020). Green hydrogen characterisation initiatives: Definitions, standards, guarantees of origin, and challenges. *Energy Policy, 138*.
57. Bloomfield, E. F. (2019). The rhetoric of energy Darwinism: Neoliberal piety and market autonomy in economic discourse. *Rhetoric Society Quarterly, 49*(4), 320–341.
58. IRENA. (2020). *Renewable Power Generation Costs in 2019*. Abu Dhabi: International Renewable Energy Agency.
59. Verma, A. (2020, December 21). Solar Tariffs Drop to a new low of Rs. 1.99/kWh in Latest Gujarat Auction. *Saur Energy International*.
60. Prasad, N. T. (2020c, November 27). New Solar Tariff Record of ₹2/kWh: Why Did it Go So Low? *Mercom India*.
61. Prasad, R. (2021, March 12). Solar tariff to rise as customs duty on gear to kick in. *ETEnergyworld*.
62. ETEnergyWorld. (2021b, Februrary 4). Budget 2021: Focus on renewable energy and Discom viability to transform power sector. *ETEnergyWorld*.
63. ETEnergyWorld. (2021d, March 16). Solar tariff could rise up to 40–45 p/unit over customs duty change. *ETEnergyWorld*. Retrieved from https://energy.economictimes.indiatimes.com/news/renewable/solar-tariff-could-rise-up-to-40-45p/unit-over-customs-duty-change/81532952
64. PTI. (2019, July 24). Government amends bidding guidelines for wind power projects. *The Economic Times*.
65. Gopal, S. (2019, July 2). Wind sector hopes for better times ahead. *Mogabay India*.
66. ETEnergyWorld. (2021c, March 12). Payment risk from discoms resurfaces for wind projects in four states: CRISIL. *ETEnergyWorld*.
67. IEA. (2020). *Covid-19 and the Resilience of Renewables*. International Energy Agency.
68. George, A. (2020, November 12). India can push up renewable energy the most after COVID-19: IEA. *Down To Earth*.
69. Pradhan, S., Ghose, D., & Shabiruddin. (2020). Present and future impact of COVID-19 in the renewable energy sector: a case study on India. *Energy Sources, Part A: Recovery, Utilization, and Environmental Effects*.
70. Chandrasekaran, K. (2020, June 20). COVID-19 impact: Renewable energy installation continues to remain slow. *ET Energyworld*.
71. Das, N., & Gambhir, A. (2020). *Rising Stakes: An Analysis of Regulatory Treatment of Renewable Electricity in Maharashtra from 2010–2020*. Prayas Energy Group.
72. Palchak, D., Cochran, J., Ehlen, A., McBennett, B., Milligan, M., Chernyakhovskiy, I., et al. (2017). *Greening the Grid: Pathways to Integrate 175 Gigawatts of Renewable Energy into India's Electric Grid, Vol. I—National Study*. National Renewable Energy Laboratory, Lawrence Berkeley National Laboratory, Power System Operation Corporation, and the United States Agency for International Development.
73. McPherson, M., & Stoll, B. (2020). Demand response for variable renewable energy integration: A proposed approach and its impacts. *Energy, 197*.
74. The New Indian Express. (2021, January 22). Solar plants: Andhra government decides not to burden Discoms. *The New Indian Express*.
75. De, A. (2020). *The Energy Sector Post COVID-19: Refueling the recovery*. KPMG.
76. Koundal, A. (2021b, March 30). Meeting 175-GW renewable energy target: Why India needs to pull up efforts. *ETEnergyWorld*. Retrieved from https://energy.economictimes.indiatimes.com/news/renewable/meeting-175-gw-renewable-energy-target-why-india-needs-to-pull-up-its-socks/81762083
77. Garg, V., & Beaton, C. (2020, October 30). Why renewable energy must play a greater role in COVID-19 economic recovery. *The Wire*.

78. Ministry of Power. (2006, January 6). *National Tariff Policy*. Retrieved March 16, 2021, from http://www.orierc.org/documents/National%20Electricity%20Tariff%20Policy.pdf
79. Prayas (Energy Group). (2021, January). *Renewable Energy Data Portal*. Retrieved March 27, 2021, from https://www.prayaspune.org/peg/re.html
80. Prateek, S. (2019a, May 17). IEX introduces green term ahead market, invites comments from stakeholders. *Mercom India*.
81. CERC. (2020, June 17). *In the Matter of Determination of Forbearance and Floor Price for the REC Framework*. Retrieved April 5, 2021, from https://www.iexindia.com/Uploads/CircularUpdate/17_07_2020Circular%20362%20REC%20Forbearance%20and%20Floor%20price.pdf
82. Financial Express. (2019, June 12). Centre, state discuss renewable energy issues. *Financial Express*. Retrieved from https://www.financialexpress.com/industry/centre-state-discuss-renewable-energy-issues/1604589/
83. Parikh, A. (2020, May 18). Karnataka extends renewable purchase obligation compliance for FY 2020 to August. *Mercom India*. Retrieved from https://mercomindia.com/karnataka-extends-rpo-compliance/
84. Government of Punjab. (2020, April 7). *Order No. 1/4/2020-EB (PR)/184*. Retrieved from Punjab Electricity Regulatory Commission: http://pserc.gov.in/pages/Order%20dated%207.4.2020-%20Direction%20to%20PSERC.pdf
85. MNRE. (n.d.). *Renewable Purchase Obligation User Manual: National Portal*. Delhi: MNRE. Retrieved March 16, 2021, from https://rpo.gov.in/
86. Sarode, J., Gambhir, A., Das, N., & Dixit, S. (2017, September). *Choosing Green: The Status and Challenges of Renewable Energy Based Open Access. Working Paper*. Retrieved October 2018, from Prayas (Energy Group): http://www.prayaspune.org/peg/publications/item/364.html
87. Singh, D. (2017, April). *Newer Challenges for Open Access in Electricity: Need for Refinements in the Regulations*. Retrieved October 2018, from Brookings India: https://www.brookings.edu/wp-content/uploads/2017/04/open-access_ds_042017.pdf
88. Rajeshwari, A. (2020, July 14). Adoption of solar through open access: Roadblocks and the road ahead. *Mercom India*.
89. Bridge to India. (2018, June). *India Solar Open Access Market*. Retrieved March 16, 2021, from https://bridgetoindia.com/report/india-open-access-report-i-june-2018/
90. Ministry of Law and Justice. (2003, May 26). *The Electricity Act, 2003*. Retrieved October 2018, from Central Electricity Regulatory Commission: http://cercind.gov.in/Act-with-amendment.pdf
91. Gambhir, A., Kokate, S., Josey, A., & Dixit, S. (2021, January 19). *Prayas (Energy Group)*. Retrieved March 15, 2021, from https://www.prayaspune.org/peg/resources/power-perspective-portal/253-renewables-moving-beyond-concessions-and-waivers.html
92. Josey, A., Dixit, S., Chitnis, A., & Gambhir, A. (2018). *Electricity distribution companies in India: Preparing for an uncertain future*. Prayas (Energy Group).
93. Forum of Regulators. (2013). *Evolving Net-metering Model Regulation for rooftop based solar PV projects*. Forum of Regulators. Retrieved from http://www.forumofregulators.gov.in/Data/HomePage/Report.pdf
94. Gambhir, A., Jethmalani, R., Sarode, J., Das, N., & Dixit, S. (2016). *India's Journey Towards 175 GW Renewables by 2022*. Prayas (Energy Group). Retrieved from http://www.indiaenvironmentportal.org.in/files/file/Indias%20Journey%20towards%20renewable%20energy.pdf
95. Das, N., Gambhir, A., Sarode, J., & Dixit, S. (2017). *India's journey towards 175 GW renewables by 2022—A July 2017 update*. Prayas (Energy Group). Retrieved from http://www.indiaenvironmentportal.org.in/files/file/India%E2%80%99s%20journey%20towards%20175%20GW%20Renewables%20by%202022.pdf
96. Vembadi, S., Das, N., & Gambhir, A. (2018). *175 GW Renewables by 2022: A September 2018 Update*. Prayas (Energy Group). Retrieved from http://www.indiaenvironmentportal.org.in/files/file/175-GW-renewables-by-2022.pdf

97. MoP. (2020, December 21). *Rights to the Electricity Consumers through "Electricity (Rights of Consumers) Rules, 2020"*. Retrieved March 17, 2021, from https://pib.gov.in/PressReleasePage.aspx?PRID=1682384
98. Prasad, N. T. (2020d, December 22). No net metering for rooftop solar systems over 10 kW: Ministry of power. *Mercom India*. Retrieved from https://mercomindia.com/no-net-metering-for-rooftop-solar/
99. Koundal, A. (2021a, March 16). Net-metering provision: Let the choice be with consumers, says industry. *ETEnergyWorld*. Retrieved from https://energy.economictimes.indiatimes.com/news/renewable/net-metering-provision-let-the-choice-be-with-consumers-says-industry/81533002
100. Singh, S. C. (2021, January 16). Examining revision of 10 kW net metering cap: Power ministry. *ETEnergyWorld*. Retrieved from https://energy.economictimes.indiatimes.com/news/renewable/examining-revision-of-10-kw-net-metering-cap-power-ministry/80298677
101. CEEW. (2020, January 15). *Streamlining Forecasting, Scheduling and DSM Regulations for Wind in India*. Retrieved March 18, 2021, from Council on Energy, Environment and Water: https://cef.ceew.in/masterclass/analysis/streamlining-forecasting-scheduling-dsm-regulations-for-wind-in-india
102. Carley, S., & Konisky, D. M. (2020). The justice and equity implications of the clean energy transition. *Nature Energy, 5*, 569–577.
103. Johnson, O. W., Han, J. Y.-C., Knight, A.-L., Mortensen, S., Aung, M. T., Boyland, M., & Resurrección, B. P. (2020). Intersectionality and energy transitions: A review of gender, social equity and low-carbon energy. *Energy Research & Social Science, 70*.
104. Climate Justice Alliance. (n.d.). *Just Transition Principles*. Retrieved November 20, 2020, from Climate Justice Alliance: https://climatejusticealliance.org/just-transition/
105. Bandyopadhyay, J., & Shiv, V. (1988). Political economy of ecology movements. *Economic & Political Weekly, 23*(24), 1223–1232.
106. Franquesa, J. (2018). *Power struggles: Dignity, value, and the renewable energy frontier in Spain*. Indiana University Press.
107. Love, T., & Garwood, A. (2013). Electrifying transitions: Power and culture in rural Cajamarca, Peru. In S. R. Sarah Strauss (Ed.), *Cultures of energy: Power, practices, technologies* (pp. 147–162). Left Coast Press Inc.
108. Zinecker, A., Gass, P., Gerasimchuk, I., Jain, P., Moerenhout, T., Oharenko, Y., et al. (2018). *Real people, real change: Strategies for just energy transitions*. International Institute for Sustainable Development.
109. Roy, A., Kuruvilla, B., & Bhardwaj, A. (2019). Energy and climate change: A just transition for Indian labour. In E. N. Dubash (Ed.), *India in a warming world: Integrating climate change and development* (pp. 284–300). Oxford University Press.
110. Roy, B., & Schaffartzik, A. (2021). Talk renewables, walk coal: The paradox of India's energy transition. *Elsevier Public Health Emergency Collection*.
111. Aggarwal, M. (2020, October 30). As India plans clean energy transition, questions remain about impact. *Mongabay India*. Retrieved from https://india.mongabay.com/2020/10/as-india-plans-clean-energy-transition-questions-remain-about-impact/
112. Stock, R., & Birkenholtz, T. (2019). The sun and the scythe: Energy dispossessions and the agrarian question of labor in solar parks. *The Journal of Peasant Studies*. https://doi.org/10.1080/03066150.2019.1683002
113. Lakhanpal, S. (2019). Contesting renewable energy in the global south: A case-study of local opposition to a wind power project in the Western Ghats of India. *Environmental Development, 30*, 51–60.
114. Rao, B. S. (2019, March 18). *How just is the transition to solar?* Retrieved March 19, 2021, from Law School Policy Review & Kautilya Society: https://lawschoolpolicyreview.com/2019/03/18/how-just-is-the-transition-to-solar/
115. Prasher, G. (2021, March 12). Solar projects on water could come at a cost to the environment, alert experts. *Mongabay India*. Retrieved from https://india.mongabay.com/2021/03/solar-projects-on-water-could-come-at-a-cost-to-the-environment-alert-experts/

116. Pratap, A., Pillai, P., & Muthu, A. (2019). *Power ahead: An assessment of the socio-economic and environmental impacts of large-scale renewable energy projects and an examination of the existing regulatory context.* Asar Social Impact Advisors Pvt. Ltd. Retrieved from https://in.boell.org/sites/default/files/2020-08/Powering-ahead_HBF_3.5.pdf
117. Dharmadhikary, S. (2020, August 19). *Solar and wind energy projects in India need an impact assessment framework.* Retrieved March 19, 2021, from Gaon Connection: https://en.gaonconnection.com/solar-and-wind-energy-projects-in-india-need-an-impact-assessment-framework/
118. Sahu, G. (2019, September 23). Whither the national green Tribunal? *Down To Earth.* Retrieved from https://www.downtoearth.org.in/blog/environment/whither-the-national-green-tribunal--66879

Chapter 7
Energy, Climate Change and Sustainable Development in India

Shailly Kedia and Nivedita Cholayil

Abstract Economies advancing in the stages of development strive for higher industrialization which requires reliable and adequate energy supply. All economic sectors including commercial, transport, residential and agriculture are dependent on different forms of energy. Attaining energy security has been one of the central pillars for countries around the world as energy is essential for supporting goals for economic growth as well as human well-being. Energy is also central to all aspects of sustainable development, and the sustainable development goals contain a dedicated goal on energy. According the Fifth Assessment Report of the Intergovernmental Panel on Climate Change, about 76% is from energy-based activities. International Energy Agency, in its report, showed that energy-related CO_2 emissions at a global level touched a historic high in 2018 [1]. As the world talks of decarbonizing and net zero goals, a re-examining of energy supply and demand is needed to critically assess narratives and policy approaches in countries considering the energy needs of countries. Using the lens of energy security and political economy, this chapter aims to critically examine recent narratives in India on energy in the context of climate change and sustainability.

Keywords Energy policy · Sustainable energy · Public policy · Climate change · Sustainable development · Environmental sustainability

7.1 Introduction

Economies advancing in the stages of development strive for higher industrialization which requires reliable and adequate energy supply. All economic sectors including commercial, transport, residential and agriculture are dependent on different forms of

S. Kedia (✉) · N. Cholayil
The Energy and Resources Institue, Darbari Seth Block, India Habitat Centre, New Delhi 110003, India
e-mail: shailly.kedia@teri.res.in

N. Cholayil
e-mail: nivedita.cholayil@teri.res.in

energy. These sectors then linked to employment, productivity and human development. For developing countries, even a small incremental rise in per capita energy use or even per capita electricity use has relatively larger increments in human development [2]. Any energy-dependent economy follows a growth or feedback hypothesis, which implies that energy and economic growth are interdependent, increasing the level of per capita energy consumption leads to higher economic growth, and vice versa [3].

When wealth is examined within the normative framework of sustainable development, intergenerational well-being is essential which takes into account resources for future generation after depletion. A way to measure this could be through energy depletion, which is the ratio of the present value of stock of energy resources, which includes coal, crude oil and natural gas, to the remaining reserve lifetime [4]. Figure 7.1 shows monetized value of energy depletion in India in absolute terms and as a percentage of gross national income (GNI). The macro-level cost of energy depletion in India has ranged from 0.35–1.64% of GNI. Natural resources also have the element of incommensurability, and market prices or even monetization of energy-related natural resources may not accurately reflect the value of such resources.

Attaining energy security has been one of the central pillars for countries around the world as energy is essential for human development and economic growth. Energy is also central to all aspects of growth and development, with the United Nations' sustainable development goals (SDGs) dedicating a goal on energy, SDG 7, which calls for ensuring access to affordable, reliable, sustainable and modern energy for all. According the Fifth Assessment Report of the Intergovernmental Panel on Climate Change (IPCC), energy-based activities such as electricity and heat production (25%), industry (21%), transportation (14%), other energy (10%) and buildings (6%) accounted for 76% of the global emissions in 2010, while the remaining 24% came from agriculture, forestry and land use-related activities [5]. Global energy-related CO_2 emissions grew reach a historic high of 33.1 Gt CO_2 in 2018 [1]. As

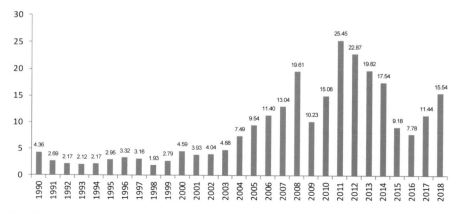

Fig. 7.1 Energy depletion for India (1990–2018) (USD). *Source* Based on World Bank [4]

the world leaders talk of decarbonizing and net zero goals, a re-examining of energy supply and demand is needed to critically assess narratives and policy approaches in countries considering the energy needs of countries. A re-examining of energy supply and demand is needed to critically assess narratives and policy approaches in countries considering the energy needs of countries. This chapter aims to critically examine recent narratives in India on energy in the context of climate change and sustainability. The following research questions (RQ) will be covered.

RQ 1: When examining of primary energy supply and energy consumption in India, what issues become relevant for sustainable development and climate action?

RQ 2: What is the role of energy security in present discourses of India?

RQ 3: Using the lens of political economy, what are the implications for energy policy in India when it comes to sustainability?

The first section is the introduction section which will be followed by energy supply and demand analysis where the supply side will cover aspects related to primary energy and fuel mix. The energy demand side will consider fuel consumed across sectors. The third section will examine the aspect of energy security since it has been an important aspect of the sustainable development discourse. This section will critically examine the recent discourse in India after the Paris Climate Change Agreement. The fourth section will be a discussion using political economy approaches. The final section will summarize the chapter and conclude.

7.2 Energy Supply and Demand: The Big Picture

7.2.1 Energy Supply

Figure 7.2 shows the energy supply mix for India as compared to the OECD and the world. One of the starkest observations is that the share coal in primary energy supply for the world has plateaued since 2000, while for OECD, it has decreased, and for India, the same has increased. This reliance on coal will remain extremely central on questions related to low carbon development, climate action and sustainability in India.

While the share of renewable energy in installed capacity for electricity generation has increased, the share of renewable energy in total primary energy supply for India is still miniscule (Fig. 7.2). Even in terms of power supply, there is a need for a careful look at the achievement of renewable energy targets.

The government has set a target of installing 175 GW renewable energy capacity, excluding large hydro, in the country by 2022. According to the latest statistics provided by the Ministry of New and Renewable Energy, as on 12 August 2021, the total installed renewable energy achieved the milestone of 100 GW making India stand at the fourth rank in terms of installed renewable energy capacity in the world [7]. India ranks at the fifth position for solar and at the fourth position in terms of

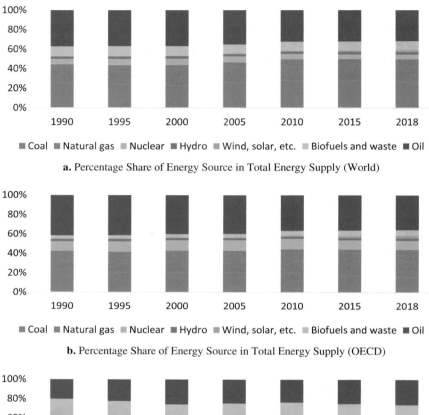

Fig. 7.2 Percentage share of energy source in total energy supply. *Source* IEA [6]

wind energy, if 50 GW under installation and 27 GW under tendering are accounted for, India is on the path of achieving the 175 GW goal [7]. India also has 450 GW target for renewable energy capacity for the timeframe of 2030. Some of the initiative taken to boost renewable energy is allowing up to 100% foreign direct investment through automatic route, notification of standard bidding guidelines to enable distribution licensee to procure solar and wind power at competitive rates, declaration of trajectory for Renewable Purchase Obligation (RPO) up to 2022, boosting new transmission lines infrastructure under Green Energy Corridor Scheme, creating Project Development Cell for attracting and facilitating investments in domestic manufacturing and doubling loan limits for renewable energy as per RBI's revised priority sector lending guidelines [8].

In 2019–20, 213.7 million metric tonnes of petroleum products were consumed, and the percentage of import dependency of oil and oil equivalent gas was 77.9% [9]. The Ministry of Petroleum and Natural Gas has identified various strategies including primarily boosting domestic production of oil and gas along with promoting alternate fuels with a view to achieve reduction in import of crude oil. These also include reforms in existing policies such as Reforms in Exploration and Licensing Policy (2019), New Hydrocarbon Exploration Licensing Policy (2016) and Discovered Small Field Policy (2016). This apart, projects include 2D Seismic Survey and Compressed Bio Gas Plants. The government has also taken several measures to switch to the use of alternative fuels like ethanol and biodiesel through Ethanol Blending in Petrol (EBP) Programme and biodiesel blending in diesel. The government has formulated a National Bio Fuel Policy 2018 to increase availability of biofuels in country [10].

India imported 248.54 million tonnes of coal in 2019–20 [11]. Some of the steps taken by the Ministry of Coal to boost production include introducing surface miners in opencast mines which not only improves operational efficiency but also tackles environmental hazards and the introduction of mass production technology in underground coal mines. The ministry's duties include undertaking policies and initiatives to encourage exploration and development of coal and lignite reserves as well as sanctioning projects. These duties are exercised through the public sector undertakings under the administrative control of the ministry such as Coal India Limited, Neyveli Lignite Corporation Limited and Singareni Collieries Company Limited. The ministry has also been promoting coal and lignite-related R&D activities through its Coal Science and Technology Plan.

In the power sector, the Ministry of Power is the administrative authority over the Electricity Act (2003) and the Energy Conservation Act (2001). The Ministry of Power is responsible for planning, policy formulation, processing of projects for investment, monitoring of the implementation of projects, training and enacting the legislation in regard to hydro and thermal power generation, transmission and distribution. According to the Central Electricity Authority (CEA), the percentage share of electricity generation from thermal power plants in the country has reduced over the years, and the share of non-fossil power has gradually increased. The share of electricity generation from thermal power plants has reduced from 80.4% in 2015–16 to 69.7% in 2020–21 until August 2020 [12].

Atomic Energy Commission, which was set up in 1958, is the apex body regarding use of nuclear energy and formulates policy of the Department of Atomic Energy in all matters concerning nuclear energy. Presently, there are twenty-two reactors with a capacity of 6780 MW in operation in the country and nine reactors, and with a total capacity of 6700 MW, about 3% of total electricity capacity is under construction [13]. India has also entered into intergovernmental agreements (IGA) for cooperation on peaceful uses of nuclear energy. Seventeen IGAs have been signed, including with European Union, United States of America, United Kingdom, Russia, Australia, Canada, Republic of Korea, Argentina, Bangladesh, Czech Republic, France, Japan, Kazakhstan, Mongolia, Namibia, Sri Lanka and Vietnam [14]. India's nuclear power share in its energy mix has remained constant with marginal increases.

7.2.2 Energy Consumption

Figure 7.3 shows the percentage share of source in total final energy consumption for the World, OECD countries and for India. In the biofuels and waste category, India still has a very high dependence on primary solid biofuels in terms of energy consumption. Primary solid biofuels are any plant matter either used directly as fuel or converted into other forms before combustion. They can include firewood, wood chips, bark, sawdust, shavings, chips, sulphite lyes (also known as black liquor), animal materials/wastes and other solid biofuels. This pattern for India stands out when compared with the rest of the world [6].

While the percentage share of solid biomass in the total energy consumption has gone down over the years (Fig. 7.3), the absolute value has not. The absolute value for the same was 130,336 ktoe in 1990 and 155,559 ktoe in 2018 [6]. While the Government of India vaunts of achieving 100% electrification in terms of household electricity connections, the stark reality remains that even more than coal, biomass such as firewood, wood chips and animal waste remains as a major source of energy consumption. 74% of the solid biofuel consumed in India is in the residential sector (Fig. 7.4). The share of biofuels and waste for India is more due to the reliance on fuelwood for cooking.

In terms of clean cooking fuel, according to the latest National Sample Survey, 67% of rural households still relied on firewood and chip as a source of cooking fuel, while 14% of urban households relied on firewood and chip as a primary energy source of cooking [15]. There is variation in terms of conceptual understanding of energy access. In this regard, two key indicators guiding policies and programmes in India and world over include access to electricity and access to clean cooking fuels.

With 99.99% of households being electrified [16], statistics from the government Websites indicate universal electrification, and the key question now pertains to reliability of power. For rural areas, the average daily power supply for the month of April 2019 ranged from 14.25 h (Jammu and Kashmir) to 24.00 h (Gujarat) [17]. This is indicative of the unreliability of electricity infrastructure in rural areas. The average number of power cuts was as high as 57 power cuts per month for Gurbarga DISCOM [18].

In terms of the percentage share of sector for India, till 2010, the share of energy consumption for the residential sector was higher than the industry sector, but in 2018, the share of industry sector in terms of energy consumption was higher than the residential sector [6]. Going by the worldwide trends, it can be expected that India's consumption in the transport sector will increase in the coming years. To promote

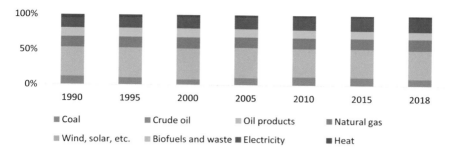

a. Percentage Share of Fuel Source in Total Final Energy Consumption (World)

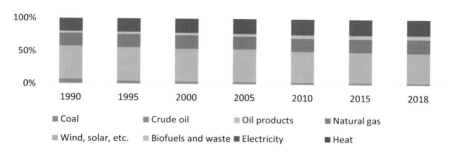

b. Percentage Share of Fuel Source in Total Final Energy Consumption (OECD)

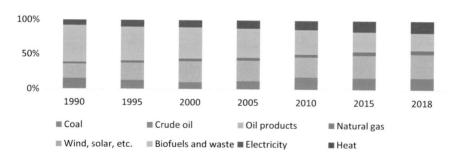

c. Percentage Share of Fuel Source in Total Final Energy Consumption (India)

Fig. 7.3 Percentage share of fuel source in total final energy consumption (India). *Source* IEA [6]

hybrid and electric vehicles, the Department of Heavy Industry has introduced a scheme called Faster Adoption and Manufacturing of (Hybrid &) Electric Vehicles in India (FAME India). Phase I of the FAME India Scheme was launched in April 2015. Phase II pf FAME India was launched in April 2019 with a total budgetary support of Rs. 10,000 crores [19].

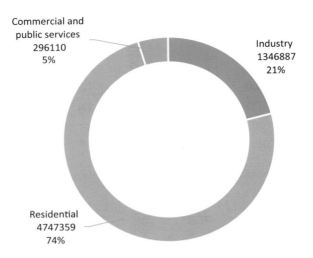

Fig. 7.4 End use sector for solid biofuel in India, 2018 (ktoe and %). *Source* IEA [6]

Additionally, the Ministry of Road Transport and Highways had notified that from April 2017, every manufacturer or importer of motor vehicles must comply with the Average Fuel Consumption Standard, as per the Energy Conservation Act, 2001 vide notification of the Government of India in the Ministry of Power number 1072 (E), dated the 23 April 2015. Moreover, the ministry has also mandated that mandated emission standard BS-VI for motor vehicles throughout the country with effect from 1 April 2020. To improve fuel efficiency, the government has notified mass emission standards for various alternate fuels such as compressed natural gas, biodiesel, ethanol (E85, E100, ED95), methanol (M15, M85 or M100), LNG, dual fuel (diesel with compressed natural gas or bio-compressed natural gas or liquefied natural gas), di methyl ether to curtail pollution and improve the fuel efficiency of the motor vehicles [20].

The agriculture sector constitutes around 18% of India's total energy consumption, with the power grid being connected to an estimated 21 million agricultural pump sets [21]. The National Energy Efficient Agriculture Pumps Programme has been launched by the Ministry of Power to help farmers replace old agricultural pumps with new 5-star rating energy efficient agricultural pumps. This programme is still in a nascent stage, and as of 15 October 2020, the numbers of pumps installed were 56,240 [21]. The Ministry of Power has made it mandatory that from 2017 onwards, all thermal plants would be based on supercritical technology as the thermal efficiency of supercritical units is typically about 2% point higher than that of subcritical units [22]. To promote energy efficiency in lighting, the Indian government launched a national light-emitting diode (LED) programme on 5 January 2015 which has two components for domestic consumers (Unnat Jyoti by Affordable LEDs for All [UJALA]) and for street lighting (Street Lighting National Programme [SLNP]).

7 Energy, Climate Change and Sustainable Development in India

To promote energy efficiency in industry, the government has identified energy-intensive sectors for which specified norms and standards for reduction in specific energy consumption (SEC) has been prescribed. The key regulatory initiative is the Perform, Achieve and Trade (PAT) scheme. In its first cycle (2012–13 to 2014–15), eight sectors comprising 478 industries were given SEC reduction targets [23]. In terms of performance in the first cycle, energy savings target was exceeded as energy savings of 8.67 million tonnes of oil equivalent (MTOE) was achieved against the target of 6.68 MTOE [23]. This was about 30% more than the target.

For energy efficiency in buildings, the Bureau of Energy Efficiency launched voluntary code, namely the Energy Conservation Building Code (ECBC) in 2007. This code establishes minimum energy standards for commercial buildings, which has a connected load of 100 kW or contract demand of 120 KVA and above [24]. On the demand side, given the energy needs for cooling, the voluntary guidelines of the Bureau of Energy Efficiency recommend that air conditioning temperature should be set at an optimal level of 24–26 °C [24].

7.3 Energy Security: Is It Still Relevant for India?

Since energy is essential for most economic activities, achieving energy security is essential. The International Energy Agency defines energy security as the 'uninterrupted availability of energy sources at an affordable price' [25]. Energy security is thus mainly concerned with energy supply responsive to the demand. According to a response to a parliament (Lok Sabha) question on energy security, at present, India is self-sufficient to meet its electricity requirements [26]. However, this statement reflects that in political discussions, the India government often conflates 'energy' with 'electricity'. In terms of energy, India is not self-sufficient. Figure 7.5 shows how the self-sufficiency of India has rapidly gone down over time.

Policy narratives in India on climate change have been clear on the aspect of meeting the goal of energy security and low carbon development. The National

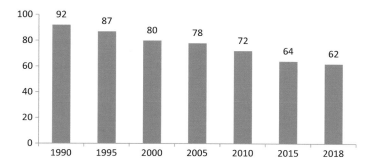

Fig. 7.5 Self-sufficiency (%) (total energy production/total energy supply), India. *Source* IEA [27]

Action Plan on Climate Change (NAPCC) communicates that the development of clean energy technologies, while primarily designed to 'promote energy security, can also generate large benefits in terms of reducing carbon emissions' [28: 13]. In its Intended Nationally Determined Contributions (INDCs), India set a target to reduce the emissions intensity of its GDP by 33–35% from its 2005 levels by 2030, along with the achievement of non-fossil fuel energy resources as the source of 40% cumulative electric power installed capacity by 2030 [29]. The role of energy measures for achieving low carbon development is key. Energy consumption in per capita terms for India remains low which is presently about 1/10th of OECD countries 1/4th of the world average [6]. In India's latest VNR, the rationale for achieving energy security is the need for low carbon development as India's contribution to world's energy-related CO_2 emission is expected to grow from 6.7 to 10.6% by 2030 [30: 75].

The Ministry of Statistics and Programme Implementation has identified two broad themes for energy indicators: one on energy use and productivity and the second on security. These indicators for India are listed in Table 7.1.

One of the key gaps in the political discourse on energy in the Indian parliament is the fact that most of the discussions, as is evident from the parliament discussions, are focused on 'electricity' and not 'primary energy'. The discourse on sustainability in India is also very much driven by diversifying fuel share in electricity generation and not in primary energy supply. At the domestic level, one of the drivers influencing

Table 7.1 Energy indicators for India, 2018

Theme	Sub-theme	Indicator	Category	Unit	Value
Use and production	Overall Use	Energy use per capita	TPES	toe/person	0.64
			TFC	toe/person	0.42
			Electricity	Kwh/person	858.85
	Overall productivity	Energy use per unit of GDP	TPES	toe/000'rupees	0.006
			TFC	toe/000'rupees	0.004
			Electricity	Kwh/000'rupees	8.64
	Supply efficiency	Efficiency of energy conversion and distribution	All	%	19.93
	Production	Reserves-to-production ratio	All	years	178.00
			Coal	years	220.00
			Lignite	years	141.00
		Resources-to-production ratio	All	years	402.00
			Crude oil	years	17.00
			Natural gas	years	41.00
			Coal	years	472.00
			Lignite	years	987.00

(continued)

Table 7.1 (continued)

Theme	Sub-theme	Indicator	Category	Unit	Value
	End use	Sectoral energy intensities	Industry	toe/000'rupees	0.009
			Agriculture	toe/000'rupees	0.001
			Transport	toe/000'rupees	0.009
		Sectoral electricity intensities	Industry	Kwh/000'rupees	13.35
			Agriculture	Kwh/000'rupees	11.33
			Transport	Kwh/000'rupees	2.43
	Diversification (fuel mix)	Fuel shares in TPES	Crude oil	%	31.26
			Natural gas	%	6.50
			Coal	%	62.92
			RE & others	%	−0.67
		Fuel share in TFC	Oil products	%	37.49
			Natural gas	%	5.69
			Coal	%	39.27
			electricity	%	17.55
		Fuel share in electricity	Thermal	%	81.91
			Nuclear	%	2.58
			Hydro	%	8.49
			RE (other than hydro)	%	7.02
Security	Imports	Net energy import dependency	Overall	%	39.21
			Crude oil	%	86.07
			Natural gas	%	44.78
			Coal	%	25.58
			Electricity	%	0.43
	Strategic fuel Stocks	Stocks of critical fuels per corresponding fuel consumption	Coal	%	6.80

Source MOSPI [31: 100]

India's position on energy and sustainability is diversification in electricity fuel mix and not diversification in primary energy mix, and in the latter, dominant share is that of fossil fuels as seen in Table 7.1. While India has ample renewable energy sources, from the perspective of demand side, technologies in sectors such as transportation are not yet ready for renewable energy. The affordability of such technologies also raises a question when it comes to a developing country like India.

7.4 Political Economy Considerations for Sustainable Energy

In the context of sustainable development and climate actions, important questions arise when examining the lens of political economy to the energy sector in India. How will India ensure access to clean and affordable energy sources such as renewable energy without leaving anyone behind, especially given the lack of reliable electricity infrastructure in the country? What are the roles of state and for-profit actors as the country transitions towards a more sustainable energy access? Since the 1990s, there has been a call for more participation from private sector business models in the energy sector approach as intergovernmental bodies such as the World Bank criticized the ability of states to take up the massive amount of expenditure on energy infrastructure [32]. This approach largely assumes that energy access is best met through market-based model, with states' only role is to foster an environment that is conducive for private sector participation [33].

Studies on developing countries across the world have provided evidence that neoliberal reforms in the energy sector over the last twenty years or so have not produced the desired results, or in some cases, have been counterproductive in terms of meeting low carbon targets, enhancing the security of supply and countering energy poverty. In fact, in some countries, there has been re-nationalization of energy sector, with those previously supporting the liberalization agenda are calling for 'reforms of reform' [34].

Within Indian context, electricity and renewable energy sector has undergone substantial liberalization, but there has always been a mismatch between electricity demand and supply, with much unreliability for those with electricity access due to unexpected outages, oscillating voltage, load shedding and erratic frequency. While renewable energy is clearly the best candidate for penetration in electricity, for a country which is increasingly concerned about climate change and energy security. However, the sustainability of this penetration will be dependent on the extent to which renewable energy costs are as competitive as coal. There are clear winners in this process: large industrial and commercial consumers who can afford to delink from distribution companies (DISCOMs) due to rising costs of grid supply and fickle supply [35]. But most importantly, there are clear losers as well: agriculture sector and low-income households. As industrial and commercial users shift towards renewable energy, state's income from electricity duties will fall, shrinking the revenue base used to subsidize the tariffs for agriculture sector and low-income households. Renewable energy also forces DISCOMs to abandon their functions of welfarism, as they come under massive pressure to become active market players.

Similarly, India's capacity in coal power has doubled in the last ten years, owing to large-scale liberalization in the sector. This resulted in large private sector players entering the sector and amass long-term profits due to ineffective support in the renewable energy sector [36]. However, currently there is a global push to move away from existing fossil-based energy systems, particularly coal assets. In order to reduce its use of coal in the primary energy mix and to meet the Paris Agreement commitment,

Indian policymakers have to achieve an ambitious target of increasing the share of natural gas to up to 15% from the present 6% by 2030. Over the years, foreign players with greater access to technology and deeper knowledge of the processes have taken a step back from investing in the sector due to differentiated pricing scheme across sectors and region. Several recommendations and appeals were made to create a uniform price for natural gas [37]. Against this background, the setting up of a natural gas exchange was welcomed. Currently, only imported gas is allowed to trade in the market. While the price of domestically produced natural gas is still determined by the government, it could eventually lead to the government stepping away from domestic natural gas as well. However, it is still too early to tell given that it is still a new platform, and it will function like a perfectly competitive market only when there are large number of buyers and sellers.

Finally, a peculiar trend which comes up in Fig. 7.3 is that India's percentage share of biofuel in the total energy consumption is much higher than the OECD countries and the world. The Ministry of Petroleum and Natural Gas introduced the National Policy on Biofuels in 2009, which was approved by the Union cabinet in 2018, largely aimed at increasing the availability of biofuels thereby substitute fossil-based fuels with biofuels. To that end, the government has set an indicative target of 20% blending of ethanol in petrol and 5% blending of biodiesel in diesel by 2030 [10]. NGOs have unsuccessfully protested biofuel policies arguing that the small farmers will only have limited benefits. If not designed properly, biofuel policy may not necessarily benefit the environment and the vulnerable communities as the policy propagates. While the Indian state claims to have converted the 'wastelands' to a more 'productive' land, including for biofuel production, a class of land brokers have emerged in the process, facilitating land-grabbing and appropriation of rents appropriation by powerful state actors as well as private companies. Those whose lands have been acquired have switched to wage labour, further marginalizing the community [38].

On the energy demand side, taking the example of clean cooking in India, an important implication of the shift towards a more flexible market-based approach has been the inability of poorer sections of the society to afford clean cook-stoves and thus were unable to take up sustainable option to cooking. Studies have shown that the market-based models for new technologies can lead to a heightened differentiation of access to finance amongst producers, fostering brokers and corruption in the process [33, 39]. On the demand side, electric vehicles (EVs) are being pushed as a measure to meet increasing demand for vehicles in a low carbon way. Studies have questioned aspects related to affordability to end consumers. One study emphasized that state backed policy measures such as tax credits and fiscal incentives will be needed to address issues related to consumer affordability for EVs [40, 41]. Another study highlighted the need for a stronger role of the government for pushing the development of charging infrastructure for EVs [42]. Thus, the role of state in facilitating accessible, affordable and accountable energy delivery systems remains key.

7.5 Conclusion

The economic growth and energy feedback hypothesis is applicable only when considering conventional GDP measurement of economic growth. Depletion of energy-related finite natural resources has implications for future generations. The discourse on sustainability at the domestic level in India is also very much driven by diversifying fuel share in electricity mix and not diversification in primary energy mix. At the domestic level, one of the drivers influencing India's position on renewable energy and sustainability has been diversification in electricity fuel mix and not diversification in primary energy mix, and in the latter, fossil fuels still including coal play an important role. Energy security still remains crucial as India relied on imports for about 40% of fuel needs in terms of primary energy. One key aspect to be highlighted is that SDGs do not consider aspects related to energy security which can be seen as a gap in the global framework for sustainable development. From the perspective of advancing discussions on energy policy in India, questions on energy policy still remain extremely crucial.

The share of biomass consumption in India is even higher than coal. In final energy consumption for 2018, coal consumption was 106,612 ktoe, while solid biofuel consumption was 155,559 ktoe [6]. This issue of high reliance on solid biomass, especially in the residential sector, in total energy consumption in India has also not received adequate attention in policy discourse. Similarly, the aspect of reliability of electricity supply especially in rural India is important as there are regions in India which still do not receive continuous electricity supply. The key question also then pertains to the quality and reliability of power. This too needs to reflect in stronger way on policy and political discourses in India.

To meet the increasing energy needs, more and more private players will play a role in the provisioning and delivery of energy services. As this takes place, the role of the state is extremely important to ensure that vulnerable communities and sectors are not left behind.

Acknowledgements The first author thanks Konrad-Adenaur-Stiftung for supporting The Energy and Resources Institute for the dialogues on energy security. The background research for the project helped in shaping this chapter.

References

1. International Energy Agency (IEA). (2019). *Global energy & CO_2 status report: The latest trends in energy and emissions in 2018*. IEA, Paris.
2. Steinberger, J. (2016). *Energising human development. Human Development Report*. United Nations Development Programme, New York.
3. Hall, C., & Klitgaard, K. (2018). *Energy and the wealth of nations: An introduction to biophysical economics*. Springer.
4. World Bank. (2020). World Bank open data. https://data.worldbank.org/ Accessed August 20, 2021.

5. Intergovernmental Panel on Climate Change. (2014). Climate Change 2014: Mitigation of climate change. In O. Edenhofer, R. Pichs-Madruga, Y. Sokona, E. Farahani, S. Kadner, K. Seyboth, A. Adler, I. Baum, S. Brunner, P. Eickemeier, B. Kriemann, J. Savolainen, S. Schlömer, C. von Stechow, T. Zwickel, J. C. Minx (Eds.), *Contribution of Working Group III to the Fifth Assessment Report of the Intergovernmental Panel on Climate Change.* Cambridge University Press.
6. International Energy Agency (IEA). (2020a). *World energy balances 2020 edition database documentation.* IEA, Paris.
7. Ministry of New and Renewable Energy (MNRE). (2021). India achieves 100 GW milestone of installed renewable energy capacity (Press Release). Press Information Bureau, New Delhi.
8. Ministry of New and Renewable Energy (MNRE). (2020b). Measures to achieve Renewable Energy objectives. Unstarred question no. 1989. Lok Sabha, New Delhi.
9. Ministry of Petroleum and Natural Gas (MOPNG). (2020). Reduction in Import of Oil. Unstarred question no. 1352. Lok Sabha, New Delhi.
10. Ministry of Petroleum and Natural Gas (MOPNG). (2018). National Policy on Biofuels. New Delhi.
11. Ministry of Coal (MOC). (2020). Coal Stock. Unstarred question no: 1835. Lok Sabha, New Delhi.
12. Ministry of Power (MOP). (2020b). Dependence on Thermal Power. Unstarred question no. 2000. Lok Sabha, New Delhi.
13. Minister of Atomic Energy (MOAE). (2020a). Investment in Atomic Energy. Unstarred question no. 579. Lok Sabha, New Delhi.
14. Minister of Atomic Energy (MOAE). (2020b). Investment in Atomic Energy. Unstarred question no. 4006. Lok Sabha, New Delhi.
15. The Energy and Resources Institute (TERI). (2019). *TERI Energy and Environment Data Diary and Yearbook.* TERI, New Delhi.
16. Ministry of Power (MOP). (2020c). Saubhagya Dashboard [Online: web]. https://saubhagya.gov.in/. Accessed October 15, 2020.
17. Ministry of Power (MOP) Ministry of Power. (2019a). Status of Rural Power Supply (April 2019). Government of India, New Delhi. https://npp.gov.in/dashBoard/rd-map-dashboard. Accessed August 18, 2021.
18. Ministry of Power (MOP) Ministry of Power. (2019b). Dashboard Data (May 2019). Government of India, New Delhi. https://urjaindia.co.in. Accessed August 18, 2021.
19. Ministry of Heavy Industries and Public Enterprises (MOHIPE). (2021). Promotion of Electric Vehicles. Lok Sabha starred question no. 398. Government of India, New Delhi.
20. Ministry of Road Transport and Highways (MORTH). (2020). Investment in Atomic Energy. Unstarred question no. 4253. Lok Sabha, New Delhi.
21. Ministry of Power (MOP). (2020d). National AgDSM Dashboard [Online: web]. http://agdsm.in/. Accessed October 15, 2020.
22. Ministry of Power (MOP). (2019c). Efficiency of Thermal Power Plants. Unstarred question no. 1767. Lok Sabha, New Delhi.
23. Ministry of Power (MOP) Ministry of Power. (2016). Energy Efficiency. Lok Sabha Question. Government of India, New Delhi.
24. Bureau of Energy Efficiency. (2017). Energy Conservation Building Code. BEE, Ministry of Power, Government of India, New Delhi.
25. International Energy Agency (IEA). (2020b). Energy security. https://www.iea.org/topics/energy-security. Accessed August, 18 2021.
26. Ministry of New and Renewable Energy (MNRE). (2020). Energy requirements per person. Unstarred question no. 781. Lok Sabha, New Delhi.
27. IEA (International Energy Agency (IEA) (2020c) IEA atlas of energy, http://energyatlas.iea.org/#!/tellmap/-297203538/1. Accessed August 18, 2021.
28. Government of India (GOI). (2008). *National Action Plan on Climate Change.* GOI, New Delhi.

29. Government of India (GOI). (2015). *Intended nationally determined contribution: Working towards climate justice.* https://www4.unfccc.int/sites/ndcstaging/PublishedDocuments/India%20First/INDIA%20INDC%20TO%20UNFCCC.pdf Accessed August 18, 2021.
30. Government of India (GOI). (2020). *India Voluntary National Review 2020.* United Nations High Level Political Forum. New Delhi.
31. Ministry of Statistics and Programme Implementation (MOSPI). (2019). *Energy Statistics 2019.* MOSPI, Government of India, New Delhi.
32. World Bank. (1991). *World development report 1991: The challenge of development.* Oxford University Press.
33. Brown, E., & Cloke, J. (2017). Energy and development: The political economy of energy choices. *Progress in Development Studies, 17*(2), vii–xiv.
34. Hall, D., van Niekerk, S., Nguyen, J., Thomas, S. (2013). *Liberalisation, privatisation and public ownership.* Public Services International (PSI). https://www.world-psi.org/sites/default/files/en_psiru_ppp_final_lux.pdf. Accessed August 18, 2021.
35. Dubash, N. K., Swain, A., Bhatia, P. (2019). The disruptive politics of Renewable Energy. *The India Forum.* https://www.theindiaforum.in/article/disruptive-politics-renewable-energy. Accessed August 18, 2021.
36. Montrone, L., Ohlendorf, N., & Chandra, R. (2021). The political economy of coal in India—Evidence from expert interviews. *Energy for Sustainable Development, 61*, 230–240.
37. Rangarajan, C. (2014). *Report of the committee on Gas Pricing.* Ministry of Petroleum and Natural Gas, New Delhi.
38. Baka, J. (2013). The Political construction of wasteland: Governmentality, land acquisition and social inequality in South India. *Development and Change, 44*(2), 409–428.
39. Cohen, S., Shirazi, S., & Curtis, T. (2017). *Can we advance social equity with shared, autonomous and electric vehicles.* Institute of Transportation Studies at the University of California.
40. Guo, S., Kontou, E. (2021). Disparities and equity issues in electric vehicles rebate allocation. *Energy Policy, 154*, 112291
41. Tseng, H. K., Wu, J. S., & Liu, X. (2013). Affordability of electric vehicles for a sustainable transport system: An economic and environmental analysis. *Energy Policy, 61*(2013), 441–447.
42. Coffman, M., Bernstein, P., & Wee, S. (2017). Electric vehicles revisited: A review of factors that affect adoption. *Transport Reviews, 37*(1), 79–93.

Part III
The Emergence of Modern Cities: Smart or Sustainable?

Chapter 8
Smart City: Sustainable City for Tackling Urban Challenges

Vinay Kandpal, Vikas Tyagi, and Harmeet Kaur

Abstract India has witnessed massive growth in its urban population over the last few decades. Governments and policymakers are facing challenges such as increasing urban population in rural areas and huge infrastructure gaps. A smart city would be a city that has facilities such as smart people, smart technologies, smart energy, smart transportation, smart IT and communication and especially smart governance. This paper aims to focus on the key issues and the challenges to develop new cities or improve the infrastructure facilities in existing cities in India, which are overpopulated and not properly managed based on an in-depth literature review of relevant studies as well as official documents of international institutions. In addition, the document also focuses on climate change issues and how to address them. Cities contribute the majority of global economic activity, energy consumption and greenhouse gas emissions. Consequently, to significantly reduce emissions, urban centres will need to consume less energy and take greater advantage of periods of intermittent availability of renewable energy. "Smart cities" should be central to achieving these goals.

Keywords Smart city · Urban development · Energy · Climate change · Renewable energy

8.1 Introduction

The rapid growth of the Indian economy has focused on physical infrastructure, social infrastructure and institutional infrastructure, since all three major areas are already in deficit. A smart city could resolve all those problems. The concept of

V. Kandpal (✉)
School of Business, Department of General Management, UPES, Dehradun, India

V. Tyagi
Department of Management, Chandigarh University, Ajitgarh, India

H. Kaur
Management Department, Jharkhand Rai University, Ranchi, India

smart city (SC) as a way to improve the quality of life of citizens has gained prominence on the agendas of policymakers. A smart city focuses on "smart governance," "smart energy," "smart environment," "smart people," "smart transport," "smart IT and smart communications," "smart buildings" and "smart life" in general. Smart applies not only to technology, but also to energy, water, transport, solid waste management and sanitation. Smart cities appeared in the literature in the late 1990s, and different approaches have been developed up to now. Until today, a smart city does not describe a city with particular attributes, but it is used to describe different cases in urban spaces: web portals that virtualize cities or city guides, knowledge bases that address local needs, agglomerations with information and communication technology (ICT) infrastructure that attract business relocation, metropolitan-wide ICT infrastructures that deliver e-services to the citizens, ubiquitous environments and recently ICT infrastructure for ecological use. The European Commission (EC) defines smart cities as places "where traditional networks and services are made more efficient with the use of digital and telecommunication technologies for the benefit of its inhabitants and business." Researchers, practitioners, business people and policymakers look at smart cities from different perspectives and most agree on a model that measures urban economy, mobility, the environment, life, people and governance. On the other hand, the ICT and construction industries insist on capitalizing the smart city, and a new market seems to be created in this area. This chapter aims to perform a literature review, discover and classify the particular schools of thought, universities and research centres as well as companies that deal with smart city domain and discover alternative approaches, models, architectures and frameworks with this regard [1].

Smart cities result in a fundamental shift in the way how citizens engage with firms where products and services are designed and delivered in a customer-centred method. While the urban population is expected to increase in the years ahead, India is facing the challenge of mass urbanization. Although the smart city is an area of opportunity for infrastructure companies and developers, this is a long-term project that will take no less than 20 years. In India, given its demographics and diversity, unique challenges and opportunities exist for developing "smarter" cities that attract increased investment, employ innovative technology, create environmentally sustainable solutions, grow operational efficiencies and amend the lives of urban citizens [2]. Many countries have already shown interest, including Japan, which is keen to develop Varanasi as a smart city, and Singapore, which has indicated Andhra Pradesh's new capital as its choice. France, the United Kingdom and the United States are just as enthusiastic. Smart cities are about using information technology to improve the efficiency and effectiveness of urban infrastructure and service delivery and to advance the sustainable development agenda. Smart cities usually involve a variety of players with diverse programs [3] (Fig. 8.1).

Essentials of Smart City: To make smart city projects in India successful, the following features are essential:

(a) **Smart Environment**: In order to create a better and healthier environment, smart cities need to be environmentally sustainable. This will not only improve

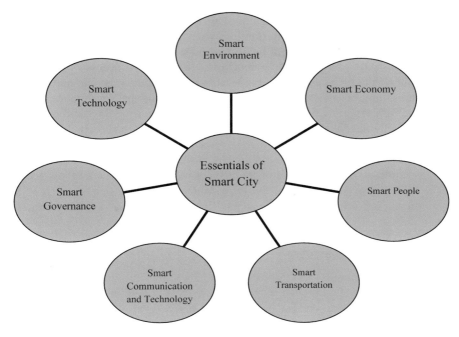

Fig. 8.1 Essentials of smart city

air quality, but also reduce waste of water, power, fuel, etc. To add further, smart environment mainly deals with

- Waste management and recycling techniques.
- Sustainable resource management for towns under energy stress.
- Combating pollution effectively and protecting the environment.

(b) **Smart Governance: Current governance structures are insufficiently participative**. People do not feel that the city is theirs. So, citizens need to be part of the decision-making process. Public participation in decision-making and transparent governance are essential to smart cities. Typically, the principle to be followed is "Governance by incentives rather than governance by execution." That would mean that people are doing the right thing because there are incentives for doing it, not because they are afraid of criminal prosecution.

(c) **Smart IT & Communications**: CISCO defines "smart cities as those who adopt scalable solutions that take advantage of information and communications technology (ICT) it increases efficiencies, reduce costs and enhance the quality of life."

The widespread use of ICTs is a necessity, and only this can ensure the exchange of information and rapid communication. The ability to shop online or book tickets online or chat online is a very powerful way to reduce the need for travel, reducing congestion, pollution and energy use. Smart communication and security solutions are key to a smart city.

(d) **Smart transport: It would be appropriate for urban transport policy to also be part of a national "urbanization policy."** It is necessary to develop policy as India's population continues to grow and more people move to the cities. India is expected to become the third largest global construction market by 2020, adding 11.5 million families each year. Smart buildings will save up to 30% of water consumption, 40% of energy consumption and cut building maintenance costs by 10–30%. About 67 percent of the rural population continues to defecate in the open, and India accounts for about 50 percent of the world's open defecation The Government of India and the World Bank have signed a USD 500 million credit for the Rural Water Supply and Sanitation (RWSS) project in the Indian states of Assam, Bihar, Jharkhand and Uttar Pradesh.

8.2 Literature Review

This conceptual paper explores how we can think of a particular city as a smart city, building on recent practices to make cities smart. A set of the common multi-dimensional components underlying the smart city concept and the core factors for a successful smart city initiative is identified by exploring current working definitions of smart city and a diversity of various conceptual relatives similar to a smart city. The paper offers strategic principles aligning to the three main dimensions (technology, people and institutions) of smart city: integration of infrastructures and technology-mediated services, social learning for strengthening human infrastructure and governance for institutional improvement and citizen engagement [4].

Ryser [5] reconnoiter the belief of "smart city" by contrasting and comparing a narrow with an eclectic understanding of "smart cities" and by placing the concept of "smart city" into the context of certain city typologies engendered over the last few decades. It considers research, debates, government policies and industry statements to analyse smart cities' contributions to existing urban policies and planning strategies.

The paper analyses the concept of an intelligent city from a critical point of view, focusing on the power/knowledge implications for the contemporary city. On the one hand, smart city policies support new ways of imagining, organizing and managing the city and its flaws; on the other, they impress a new moral order on the city by introducing specific technical parameters to distinguish between the "good" and "bad" city. The Smart City discourse can therefore be a powerful tool for the production of docile subjects and mechanisms of political legitimization. The paper is largely based on theoretical reflections and uses smart city policy in Italy as a case study. The paper analyses how the smart city discourse proposed by the European Union has been reclassified to produce new visions of the "good city" and the role of private actors and citizens in the management of urban development [6].

This paper reviews the factors which differentiate policies for the development of smart cities, to provide a clear view of the strategic choices that come forth when mapping out such a strategy. The paper begins with a review and categorization of four strategic choices with space reference, based on recent literature and the experience of smart cities. The pros and cons of each strategic option are presented. In the second part of the paper, the previous selections are illustrated by cases of smart city strategies from around the world. The third part of the paper contains recommendations for the development of smart cities based on the combined conclusions of the preceding parts. The paper concludes with a discussion of the ideas that have been provided and recommendations for future fields of research [7].

Egenhofer [8] highlights smart cities' goal of accelerating investments and inventions. The financial system is critical to the existence of the market, and new models and markets can be created through new financial mechanisms. The creation of the financial system requires attracting potential investors by introducing innovative financial products depending upon the risk, return requirements of the investors, monetization of indirect economic benefits and encouraging citizens to participate in the funding process. The paper suggests the creation of new financial models such as smart city (municipal) project obligations and the involvement of local municipal governments in planning the funding process. The paper also identifies barriers and potential solutions for funding smart cities. The suggested measures for attracting the required capital for investment are inviting the funds from special institutions like pension funds, reducing the real and perceived risk of investment, reduction of transaction cost and developing of off-balance sheet investment mechanism.

The SC policy in India is one of the very few instances globally where a central government has taken it upon itself to shepherd local governments through the process of becoming "smart", a model that relies on the integrative capabilities of the implementing agency [9] while allowing for a bottom-up approach to identifying context-sensitive solutions to urban needs. Research in this fast-emerging interdisciplinary area highlights that smart cities attempt to fulfil multiple urban development objectives, solve a variety of problems typical to modern cities and achieve many different outcomes [10, 11].

In developed countries, the challenge is to deal with a lack of infrastructure, such as drinking water, sanitation and public transportation systems. There are growing needs for a renewal of infrastructures, such as water, transportation and energy systems, which deteriorate over time, and the related increasing challenges regarding the sustainability of the systems. Drivers of change include lower costs, higher levels of efficiency, better response to natural disasters (resiliency), ability to provide good service, etc. Cities, regulators and operators focus on enhancing innovation and the development of smart cities and infrastructure. Public–private partnerships (PPPs) have been at the forefront of infrastructure development and management, but there are questions about existing models that can be applied to smart infrastructure" [12].

Lim and Maglio [13] identified 12 application areas related to smart cities from a textual analysis of 1234 news articles; these are "smart device," "smart environment," "smart home," "smart energy," "smart building," "smart transportation," "smart logistics," "smart farming," "smart security," "smart health," "smart hospitality" and "smart education." They form a hierarchy of smart cities. In smart cities, local resources, government, companies, citizens and visitors are connected by smart devices and smart environments, key resources that facilitate the collection of data from the resources and stakeholders and the delivery of various smart services to the stakeholders, such as smart energy, transportation and health services. [14] argue that smart cities offer opportunities for multi-national companies to innovate and develop new technologies.

8.3 Objective of Study

This paper is an attempt to discuss and state the issues involved in the flagship program of Prime Minister of India Mr. Narendra Modi and the challenges involved in the successful implementation of strategies as far as smart technology, smart environment, smart transportation, smart buildings and smart people.

8.4 Issues Related to Developing Smart City Projects

Under the flagship "Safe City" project, the Union Ministry proposes USD 333 million to make seven big cities (Delhi, Mumbai, Kolkata, Chennai, Ahmedabad, Bangalore and Hyderabad) to centre on technological progress rather than the workforce. The Ministry of Urban Development plans to invest over 20 billion dollars in metro projects in the coming years. The Government of India has approved a USD 4.13 billion plan to spur electric and hybrid vehicle production by laying out an ambitious target of 6 million vehicles by 2020. Electric vehicle charging stations are expected in all urban areas and along all national and national roadways by 2027. India has invested $1.2 billion so far and looks forward to attracting more funding from private and foreign investors. Developing a new or greenfield smart city with a target population of 5 to 10 lakhs is likely to require financial investment ranging between INR 75,000 and 150,000 crores and may require 8–10 years for implementation.

8.5 Funding for Smart Cities Projects in India

India plans to create 100 new smart cities and develop modern satellite cities around existing cities through the smart cities programme. Investments of approximately $1.2 trillion will be required over the next 20 years in areas such as marine transportation, energy and public safety to build smart cities in India. A robust capital market, innovative business model, a sound business environment, public private partnership and world-class investment quality projects are essentials for smart cities. Those are the foundations of the financial infrastructure upon which a smart city rests. The risk–return profile of a smart city investment in India is unique: for both government and investors, strong and continued master planning is the key to the dynamic management of both potential risks and opportunities. It is expected that a large part of the financing for smart cities will have to come from the private sector with the states/cities and the central government is only supplementing that effort.

To attract funding, policymakers should step up their risk mitigation efforts to make their smart city competitive not only compared to other Indian urban centres, but also in relation to globally comparable investment opportunities. Tools available for this include planning, credit enhancement, tax incentives, concession agreements and upgrade of reporting and data management systems. Above all, policymakers should look to ensure a coherent, predictable and transparent business climate for domestic and global investors with regular and meaningful dialogue with industry. $1.2 billion was allocated to smart cities, and FDI standards eased the $83 million allocated to the Digital India Initiative. PPP financing model would be used to improve and modernize infrastructure in urban areas. Smart city expects employment to increase by 10–15%. The Indian government and the World Bank signed a $236 million deal to reduce the risk of disaster in the coastal villages of Tamil Nadu and Pondicherry. Power Grid Corporation of India Ltd plans to invest $26 billion U.S. over the next five years (2012–2017).

8.6 Debt Markets

The need to develop infrastructure in India is undoubtedly enormous. However, the sector faces many fundamental challenges, including the need for new long-term investors to be involved in project financing. To date, debt financing in India has been largely managed by banks, which—with significant infrastructure assets already in place—are quickly approaching the limit of their debt. Therefore, the market is turning to pension funds and life insurance companies that are well capitalized and seeking long-term returns. U.S. life insurance and pension funds are enthusiastic about expanding their presence in India.

Municipal bond markets have also remained under-developed in India and have a long road ahead. The Ahmedabad Municipal Corporation was a pioneer, although comparable structures are rare. In the past two years, the Government of India has proposed several structures intended to mobilize debt financing, including a regulatory framework from the Reserve Bank of India (RBI) for a non-bank finance company (NBFC) which issues bonds, and one regulated by the Securities and Exchange Board of India (SEBI) for a trust structure which issues listed shares in a style similar to a mutual fund. Both the NBFC and Mutual Fund IDF routes are designated to free banks from the aforementioned asset-liability mismatch and allow for increasing lending.

8.7 Public–Private Partnerships

Smart city initiatives often rely on various types of public–private partnerships (PPPs) as infrastructure, particularly through the formal alignment of government and industry capacities, technology, assets and capital. The key to a successful PPP is the ability for both parties to prosper within the agreement, which is frequently seen with partnerships that have strong governance, realistic objectives, dedicated income streams and a manageable degree of risk equitably shared between the parties.

The development of a smart city has two components: infrastructure (communication, power, roads, sanitation, water assets, etc.) and real estate. In India, foreign ownership and investment rules for infrastructure and real estate differ. The government should consider classifying most smart city development as "infrastructure" to start with, such that for smart city development, the same ownership, financing and taxation regulations apply to both parts of the development. The nature of the concession agreement between the government and the developer is another key area to PPP success. India has sufficient experience to reach such agreements for airports and highways. The concession agreement between the government and developers must deliver a well-defined scope of the project and proper tolling agreements if the smart city is to be developed on a "Build-Operate-Transfer" (BOT) basis along with private sector actors. To reduce the coordination burden between many stakeholders, a smart city's concession agreements should consider both infrastructure and real estate development on a composite (part of the same development) basis rather than separately.

8.8 Viability Gap Funding (VGF)

The Government of India has launched a Viability Gap Funding (VGF) Scheme to raise the fiscal viability of competitively bid infrastructure projects which are warranted by economic returns, but do not exceed the standard threshold of financial returns. Under the scheme, the government provides grant assistance up to 20% of the capital cost of the project, with an additional grant of up to 20% of the project costs that can also be provided by the sponsoring authority. The contractual documents, delimiting risks, responsibility and performance standards have been designed, and a 10-year tax holiday is available for private companies investing in Indian infrastructure. The Government of India has also set up independent regulatory authorities to balance the interests of all stakeholders.

8.9 Challenges for India's Smart Cities Projects

The High Power Expert Committee on Investment Estimates in Urban Infrastructure has assessed a Per Capita Investment Cost (PCIC) of $685 for a 20 year period. The total estimated investment need for the smart city is $113 billion over 20 years (with an annual increase of 10% from 2009–10 to 2014–15).

Land acquisition, foreign direct investment and other issues have yet to be addressed. The prospect of large amounts of private sector money, both domestic and foreign, is going to be a challenge. These concerns imply that many projects may not be commercially viable at first. The failure of a PPP is often due to a lack of realistic targets, financial management, project governance and equal risk management.

Institutions that help cities manage electricity, water, waste, traffic flows, municipal operations and services are becoming more complex and costly. While return on investment can be attractive, the complexities often make it difficult for cities to get their smart city projects off the ground.

Developing smart cities in India requires addressing challenges related to political alliance, funding and stakeholder management. Greater harmonization between and within governmental bodies will be necessary. An investment policy and environment suitable for private investors is imperative. Consideration should also be given to ensuring that all stakeholders are involved in the decision-making process. There will be a need for clear lines of accountability.

Building new cities and upgrading existing ones poses a number of challenges in terms of integrated planning, policy alignment, financing and stakeholder management. An enabling policy framework and better harmonization of government organizations at all levels will be necessary.

Successful implementation of smart city solutions needs effective horizontal and vertical coordination between various institutions involving institutions providing various municipal amenities as well as effective coordination between central government (MoUD), state government as well as local government agencies on various issues related to financing, sharing of best practices and sharing of service delivery processes.

Building new ones is easier than turning the old ones into smart cities. However, it is also imperative to modernize cities, and it takes two to three decades to do so. In such instances, work should be undertaken at the domain level. This is still an ongoing job as you can still continue to update.

Other challenges for India include the integration of technology with law enforcement. There is no point in installing high-tech traffic lights if their implementation cannot be enforced. India will also need to find ways to encourage private investment for the infrastructure necessary for a smart city.

8.10 Conclusion

The concept of smart city is increasingly considered a new paradigm of sustainable growth. India's smart cities programme hopes to revolutionize urban life and improve the quality of life of the Indian urban population. A smart city would require a smart economy, smart people, smart organization, smart communications, smart engineering, smart transit, a fresh environment and a healthy lifestyle. However, with mass migration leading to basic problems, such as water shortages and overcrowding, the speed at which these cities will be developed will be key. Several initiatives are under way by the Indian government to convert 100 cities into smart cities. The government will actively use the PPP channel and promote FDI for the effective implementation of the smart cities project in India. The government is focusing on promoting public–private partnership (PPP) for the successful implementation of the smart city project in India. The financial services and IT sectors are on the government's priority list to attract investment from major companies like Cisco, EMC, GE, IBM, Bajaj, etc., over the coming years. A few of the major companies that are currently involved in the project planning of these cities include Halcrow, Synoate, Knight Frank and AECOM India. Some of the main firms currently involved in project planning for these cities are Halcrow, Synoate, Knight Frank and AECOM India. Building a smart city is not just about building physical infrastructure—roads, clean water, electricity and transportation. It is desired that public–private partnerships (PPP) will deliver, but the mechanism appears to require a lot of plucking for it to work, a fact recognized in the recent budget. The big challenge will be to create cities that are self-sustaining, that create jobs, that use resources wisely and that train people. The idea should be to employ cities in the service of the masses. India must now make an important decision in the context of developing smart cities. It must decide whether it wants to build new towns or upgrade existing ones. There is an agreement among the majority of the people in the community that state government and municipal bodies should play a vital role in fostering smart cities, which focus on improving quality of life by integrating technology with the built environment.

References

1. Anthopoulos, L. G. (2015). Understanding the smart city domain: A literature review. In M. P. Rodríguez-Bolívar (Ed.), *Transforming city governments for successful smart cities* (pp. 9–21). Springer International Publishing.
2. Li, F., Nucciarelli, A., Roden, S., & Graham, G. (2016). How smart cities transform operations models: A new research agenda for operations management in the digital economy. *Production Planning & Control, 27*(6), 514–528. https://doi.org/10.1080/09537287.2016.1147096
3. Seetharaman, A., Cranefield, J., & Chakravatry, S. (2019). Making Indian cities smart: Framing incongruencies and reconciliation. In *Smart Cities in India: Framing Incongruencies and Reconciliation, Fortieth International Conference on Information Systems, Munich* (pp. 1–17).

4. Neirotti, P., De Marco, A., Cagliano, A. C., Mangano, G., & Scorrano, F. (2014). Current trends in smart city initiatives: Some stylised facts. *Cities, 38*, 25–36. https://doi.org/10.1016/j.cities.2013.12.010
5. Ryser, J. (2014). Planning smart cities … sustainable, healthy, liveable, creative cities … or just planning cities? *Real Corp 2014 Plan it Smart* (vol. 8, no. May, pp. 447–456), [Online]. Available: http://conference.corp.at/archive/CORP2014_115.pdf
6. Vanolo, A. (2013). Smartmentality: The smart city as disciplinary strategy. *Urban Studies, 51*(5), 883–898. https://doi.org/10.1177/0042098013494427
7. Angelidou, M. (2014). Smart city policies: A spatial approach. *Cities, 41*, S3–S11. https://doi.org/10.1016/j.cities.2014.06.007
8. Egenhofer, C., et al. (2013). Smart Cities Stakeholder Platform—10 Year Rolling Agenda EU. 2013. [Online]. Available: https://www.yumpu.com/en/document/read/29687975/10-year-rolling-agenda-smart-cities-stakeholder-platform
9. Praharaj, S., & Han, H. (2019). Building a typology of the 100 smart cities in India. *Smart Sustainable Built Environment, 8*(5), 400–414. https://doi.org/10.1108/SASBE-04-2019-0056
10. Chourabi, H., et al. (2012) Understanding smart cities: An integrative framework. In *45th Hawaii International Conference on System Sciences* (pp. 2289–2297). https://doi.org/10.1109/HICSS.2012.615.
11. Nam, T., & Pardo, T. A. (2011). Conceptualizing smart city with dimensions of technology, people, and institutions. In *Proceedings of the 12th Annual International Digital Government Research Conference: Digital Government Innovation in Challenging Times* (pp. 282–291). https://doi.org/10.1145/2037556.2037602.
12. Cruz, S. (2017). Reforming traditional PPP models to cope with the challenges of smart cities. *Competency & Emotional Intelligence Quarterly, 18*(1–2), 94–114.
13. Lim, C., & Maglio, P. (2018). Data-driven understanding of smart service systems through text mining. *Service Science, 10*, 154–180. https://doi.org/10.1287/serv.2018.0208
14. van den Buuse, D., & Kolk, A. (2019). An exploration of smart city approaches by international ICT firms. *Technological Forecasting Social Change, 142*(May), 220–234. https://doi.org/10.1016/j.techfore.2018.07.029

Chapter 9
Electric Mobility and Electric Vehicles Management in India

Shikha Juyal

Abstract Mobility allows people to access various needs of their lives including jobs, education, health care and other services. India has potential to achieve electric mobility future by utilising existing conditions, government programmes and policies. To upscale adoption of electric vehicles and for its management in India, issues such as charging infrastructure, research & development, financing of electric vehicles, battery and cell manufacturing, proper regulatory framework, fiscal and non-fiscal incentives, availability of power and its infrastructure, consumer awareness need to be addressed immediately. In this backdrop, this paper initially highlights the vision and opportunity of electric vehicles in India and then tries to explain that the electric mobility pathway would provide clean, low cost mobility, create new jobs, reduce oil imports, improve health of people and would have positive economic impact. The paper highlights the policies and number of incentives provided by Government of India. In the end, the paper mentions the challenges which need to be addressed to boost adoption of electric vehicles in future.

Keywords Electric mobility · Electric vechicle management · Initiatives · Incentives · Challenges

9.1 Introduction

The transportation sector, driven almost entirely on fossil fuels in the form of petroleum products, has a large environmental footprint and linked negative externalities. Not only does the sector impact local air quality and global challenges of climate, but it also has implications on human health and biodiversity.

Declaration: The views and analysis expressed in the chapter are personally those of the author. They do not reflect the views of NITI Aayog and IGIDR. NITI Aayog and IGIDR do not guarantee the accuracy of data included in the publication nor does it accept any responsibility for consequences of its use.

S. Juyal (✉)
NITI Aayog, New Delhi, India

In addition to greenhouse gas (GHG) emissions, the sector is also responsible to large extent for emissions of pollutants such as CO, SO_x, NO_x and particulate matter, all of which directly affect human health and biodiversity. The uncontrolled growth of automobiles has led to the transport sector becoming one of the key sources responsible for degrading local air quality and pollution. According to the study on 'Air Pollution and Green House Gases in Delhi' by IIT, Kanpur in 2016, [1] vehicles contribute significantly to PM10 and PM 2.5.

As per IQ Air in 2020 [2] among the world's most polluted countries, India was ranked third in terms of PM 2.5. As per World Air Quality Report (2020), [3] India continues to dominate annual PM 2.5 rankings by city—22 of the top 30 most polluted cities globally are located in India. In Paris Agreement, India has committed to reduce the emissions intensity of GDP by 33–35% by 2030 below 2005 levels. In order to meet the global commitment and mitigate adverse impact of the automobiles, the Government of India is keen to shift towards electric vehicles. Therefore, it is important to introduce alternative modes of transport with rapid urbanisation, economic growth, increase in travel demand, climate change and energy security. Electric mobility is a viable option to address these challenges.

As per Invest India, [4] automobile sector contributes to 49% of India's manufacturing GDP and 7.1% of India's GDP. The second Automotive Mission Plan (AMP) [5] outlines the plan to elevate the automotive industry to world class levels. India is expected to be the world's third-largest automotive market in terms of volume by 2026. The industry currently manufactures 26 million vehicles including passenger vehicles, commercial vehicles, three wheelers, two wheelers and quadricycles in April–March 2020, of which 4.7 million are exported. India holds a strong position in the international heavy vehicles arena as it is the largest tractor manufacturer, second-largest bus manufacturer and third-largest heavy trucks manufacturer in the world. The electrical vehicles (EV) market is expected to grow at CAGR of 44% between 2020 and 2027 and is expected to hit 6.34 million-unit annual sales by 2027. The EV industry will create five crore direct and indirect jobs by 2030. A market size of $50 billion for the financing of EVs in 2030 has been identified—about 80% of the current size of India's retail vehicle finance industry, worth $60 billion today. India's passenger vehicle industry is expected to post a growth of 22–25% in financial year 2022. The electric vehicle market in India is expected to be valued at $2 billion by 2023. With battery costs declining faster than anticipated, electric vehicles economics and five-year TCO would become favourable.

Mobility is one of the most critical inputs required for development. It allows people to access various needs of their lives including jobs, education, health care and other services. India has low vehicle per 1000 population and has opportunity to move towards electric mobility. The shift towards electric mobility would provide clean, low cost mobility, provide new jobs, reduce oil imports, improve health of people and would have positive economic impact.

As per NITI Aayog and World Energy Council [6] report due to climate change, advances in renewable energy, urbanisation, data capture and analysis, battery chemistry and energy security are circumstances which have led electric mobility to enter the mass market.

As per Viswanathan and Sripad [7], 20 per cent of CO_2 emissions and 30 per cent of particulate emissions in India are caused by motorised two-wheelers. Therefore, under FAME, they have considered the need to electrify the motorised two-wheeler.

IEA (2021) [8] report mentioned that by end of 2020, there were 10 million electric cars. Electric car registrations increased by 41% in 2020, despite pandemic and around 3 million electric cars were sold globally (a 4.6% sales share). For the first time, Europe overtook China in terms of world's largest electric vehicle (EV) market. The global EV fleet reaches 230 million vehicles in 2030 (excluding two/three-wheelers), a stock share of 12% in Sustainable Development Scenario. The cost of electric vehicles would be down in future due to advances in battery technology and mass manufacturing.

NITI Aayog and Rocky Mountain Institute [9] report mentions that there is need to overcome the key barriers related to electric vehicles adoption including technology cost, infrastructure availability and consumer behaviour. Financing is a hurdle for India's electric mobility transition. The end-users face a range of challenges like high interest and insurance rates apply to retail loans, loan-to-value ratios are low and specialised finance options are limited. The quantum of finance required for electric vehicles adoption scenario is considerable. Between 2020 and 2030, the estimated cumulative capital cost of the country's EV transition will be INR 19.7 lakh crore (USD266 billion) across vehicles, electric vehicle supply equipment (EVSE) and batteries (including replacements). The estimated size of the annual EV finance market will be INR 3.7 lakh crore (USD50 billion) in 2030. There are six targeted instruments and four ecosystem enablers that financial institutions, the EV sector and the government can adopt to mobilise the capital and financing associated with India's EV transition. The targeted instruments are priority sector lending; interest rate subvention; product guarantees and warranties; risk-sharing mechanism (government and multilateral-led); risk-sharing mechanism (fleet operator-led) and secondary market development. The ecosystem enablers are digital lending; business model innovation; fleet and aggregator electrification targets and open data repository for EVs. Finally, innovative procurement and leasing initiatives that lead to early deployments at scale can help prove the techno-economic viability of electric vehicles and increase supply chain investments. The report mentions that supporting the design of effective financing solutions can help unlock the capital needed for India's EV transition.

9.2 Vision and Opportunity of Electric Vehicles in India

By 2026 as per Automotive Mission Plan (AMP) 2030 [5] vision, the Indian automotive industry will be among top three of the world in engineering, manufacture and export of vehicles and auto components. NITI Aayog and Rocky Mountain Institute [10] report highlights that by 2030 in transformative scenario, the percentage of electric vehicle would be 40% for two-wheeler, 100% for three-wheeler, 40% BEV for four-wheeler (personal), 100% BEV for four-wheeler (commercial) and 100%

for public transit. By persuing shared, connected and electric mobility, India can save 64% of anticipated passenger road-based, energy demand and 37% of carbon emissions in 2030. This would save Rs 3.9 lakh crore or USD ~60 billion. The report highlighted list of actionable solutions.

Stanley [11] report mentions that by 2030, India would become global leader in shared mobility. According to Morgan Stanley expectations by 2030, India will have 35% of miles shared by 2030 and 30% of EV penetration. China will have 30% of miles shared by 2030 and 30% of electric vehicles penetration.

Bloomberg NEF's 2020 Battery Price Survey [12] predicts that by 2023, average pack prices will be $101/kWh. The drop in battery prices could provide necessary thrust for high uptake of electric vehicles (EVs) by bringing them at par with internal combustion engine (ICE) vehicles.

ICCT [13] mentioned that on 11 June 2021, Department of Heavy Industries made some very encouraging modifications to the FAME-II scheme. Purchase incentives for electric two-wheelers (E2W) were increased by 50% to INR 15,000 per kWh of battery capacity. The limit on this incentive was also relaxed from 20% of the ex-showroom price to 40% of the ex-showroom price. Following the FAME-II subsidy revision announcement, manufacturers were also quick to announce retail price cuts for the shorter-range models that currently dominate the market. The latest revised incentives also make an already attractive TCO proposition even more alluring for consumers. Going back to the mid-range example, both 5-year and 10-year TCO parity was already achieved before the additional FAME-II incentives. The additional incentives combined with state-level incentives, which include direct incentives and a full road tax-waiver, make for a very compelling case by lowering electric two-wheeler TCO below conventional models.

Transition to electric vehicles can provide benefits like increase in energy security, air pollution and greenhouse gas emissions reduction and industrial development. As on 2 August, 2021, total registered electric vehicles in India were 730,237 as per e-Vahan portal. The year-wise electric vehicles sales trend in India is shown in Fig. 9.1. The total EV registered has been increasing from 2014–15 to 2019–20 in India, but due to pandemic, little slowdown was seen in 2020–21. But overall trend shows that the number of electric vehicles registered has been increasing. The electric vehicle addition was highest for three-wheeler followed by two-wheeler.

As per Society of Manufacturers of Electric Vehicles (SMEV), data published in autocar Website, [15] the electric vehicle sales in India in financial year 2021 were 143,837 for two-wheelers, 88,378 for three-wheelers, 5905 for cars, and the total was 238,120. In financial year 2020 was 152,000 for two-wheelers, 1,40,683 for three-wheelers, 2814 for cars, and the total was 2,95,497. In financial year 2021, electric two-wheeler sales down by 5.37%, three-wheeler sales down by 59% and electric passenger vehicle sales up 110 per cent. The difference in vehicle data available in e-vahan and with Society of Manufacturers of Electric Vehicles could be due to number of registered electric vehicles in India.

9 Electric Mobility and Electric Vehicles Management in India

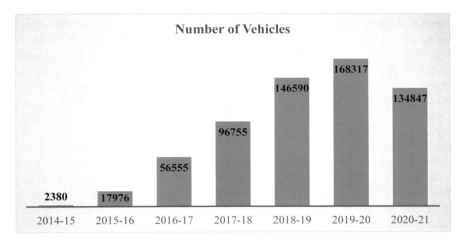

Fig. 9.1 Year-wise registered electric vehicles sales. *Source* https://vahan.parivahan.gov.in/vahan4 dashboard/vahan/view/reportview.xhtml (accessed on 2 August, 2021). This was as per e-vahan portal and for Electric (BOV) [14]

9.3 Initiatives and Incentives by Government of India

The Government of India has undertaken many initiatives to promote adoption and manufacturing of electric vehicles in India.

9.3.1 National Electric Mobility Mission Plan (NEMMP) 2020 and Faster Adoption and Manufacturing of Hybrid & Electric Vehicles in India (FAME India) Scheme

National Electric Mobility Mission Plan 2020 [16, 17] document provides vision and roadmap for electric vehicle adoption and manufacturing. It was designed to increase national fuel security, to provide affordable and environmentally friendly transportation and to enable automobile industry to achieve leadership in manufacturing. As part of National Electric Mobility Mission Plan 2020, Faster Adoption and Manufacturing of (Hybrid &) Electric Vehicles in India (FAME India) [18] Scheme was formulated in year 2015 by Department of Heavy Industry to promote manufacturing of electric vehicles in India by providing incentives on purchase of electric vehicles and for establishing charging infrastructure. Department of Heavy Industry (DHI) is nodal department for Faster Adoption and Manufacturing of Hybrid & Electric

Vehicles in India (FAME India) Scheme since 1 April 2015, Phase-I of the scheme was extended from time to time, and the last extension was allowed till 31 March 2019. Currently, Phase-II of FAME India Scheme is implemented from 1 April, 2019 for a period of 3 years with a total budgetary support of Rs. 10,000 crores. The Phase-II of FAME India Scheme will focus on electrification of public and shared transport and aim to provide demand incentive to approximate 7000 electric buses, 5 lakh electric three-wheelers, 55,000 electric four-wheeler passenger cars and 10 lakh electric two-wheelers. The creation of charging infrastructure is also supported under the scheme.

About 2.8 lakh hybrid and electric vehicles were supported in first phase of FAME Scheme by providing demand incentives of Rs. 359 crore. Department of Heavy Industry sanctioned 425 electric and hybrid buses for various cities with total cost of Rs. 300 crores and 500 charging stations/infrastructure for Rs. 43 crore (approx.) under Phase-I of FAME India Scheme. Under Phase-II of FAME India Scheme till 26.02.2020, 14,160 electric vehicles have been supported amounting Rs.50 crore, and 5595 electrical buses have been sanctioned to various State/City Transport Undertakings of around Rs. 2800 crore. DHI has also sanctioned 2636 charging stations of about Rs 500 crore (approx.) in 62 cities across 24 States/UTs.

Present status of FAME as on 6 September, 2021: As per DHI dashboard, under FAME-II, the total number of vehicles sold is 111,936, the fuel saved (in litres) is 26,010,704, and CO_2 reduction (in kg) is 59,195,734. Figure 9.2 shows number of vehicles sold state wise.

The total number of electric vehicle sold in India under FAME India Scheme as of August 2021 is 3.87 lakh electric vehicles and 6740 electric buses.

9.3.2 National Mission on Transformative Mobility and Battery Storage

To drive clean, connected, shared, sustainable and holistic mobility initiatives, the Government of India has set up National Mission on Transformative Mobility and Battery Storage [19, 20]. The role of mission is as follows:

- To recommend and drive the strategies for transformative mobility and Phased Manufacturing Programmes for Electric Vehicles, Electric Vehicle Components and Batteries.
- To launch Phased Manufacturing Program (PMP) to localise production across the entire electric vehicle value chain.
- To finalise the details of the value addition that can be achieved with each phase of localisation with a clear Make in India strategy for the electric vehicle components as well as battery.
- To coordinate with key stakeholders in ministries/departments and the states to integrate various initiatives to transform mobility in India.

9 Electric Mobility and Electric Vehicles Management in India   165

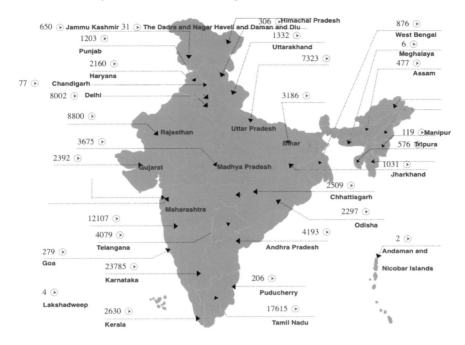

Fig. 9.2 Present status of number of vehicles sold state wise as per DHI under FAME 2 [18]. *Source* https://fame2.heavyindustry.gov.in/ accessed on 6 September 2021

9.3.3 Phased Manufacturing Programme (PMP)

To support setting of few large scale, export-competitive integrated batteries and cell manufacturing Giga plants in India, Phased Manufacturing Programme was launched which is valid for 5 years till 2024. It would localise production across electric vehicles value chain [21].

9.3.4 Production-Linked Incentive (PLI) Scheme for Manufacturing of Advanced Cell Chemistry (ACC) Battery

PLI scheme will enhance globalisation and make Indian automobile industry more competitive. It has been introduced in ten key sectors for enhancing India's manufacturing capabilities and enhancing exports. PLI scheme for manufacturing of Advanced Cell Chemistry (ACC) battery has been approved with total outlay of INR 18,100 crores for first 5 years, i.e. from 2022 to 2026. It would encourage domestic manufacturers, reduce import dependence and make India leader in electric vehicles. Through this scheme, the Government of India would be able to realise goal of Atmanirbhar Bharat. The scheme intends to optimally incentivize potential investors, both domestic and overseas, to set-up Giga-scale ACC manufacturing facilities with emphasis on maximum value addition and quality output and achieving pre-committed capacity level within a pre-defined time period. The net saving would be Rs. 200,000 crore to Rs. 250,000 crore on account of oil import bill reduction. The Advanced Cell Chemistry (ACC) programme will help in reduction of greenhouse gas (GHG) emissions as will facilitate demand for electric vehicles which is less polluting.

9.3.5 Production-Linked Incentive (PLI) Scheme for Automobile and Auto Components

Production-Linked Incentive (PLI) Scheme in the Automobile and Auto Components sectors was introduced to enhance manufacturing capabilities and exports, and the approved outlay is Rs. 57,042 crore over a five-year period.

The battery waste management rules have been amended by Ministry of Environment, Forest and Climate Change (MoEFCC) to include lithium ion batteries under the ambit of 'Extended Producer Responsibility'. Department of Heavy Industry (DHI) had floated Expression of Interest (EoI) for deployment of EV charging stations on the highways, across the country. Ministry of Road Transport and Highways (MoRTH) has allowed registration of vehicles without batteries, and this would promote alternate business models like battery swapping. MoRTH has removed mandatory requirement of spare tire in cars, and this would provide extra space for electric vehicle batteries in vehicles.

9.3.6 Charging Stations

To accelerate electric vehicle adoption, it is important to have adequate charging infrastructure. In this regard, Ministry of Power has issued 'Charging Infrastructure for Electric Vehicles—Guidelines and Standards' defining the roles and responsibilities of various stakeholders. A policy on charging infrastructure has been issued, and this would permits private charging at residences/offices where tariff for supply of electricity to electric vehicle charging station shall not be more than the average cost of supply plus 15%. Ministry of Power has issued a notification clarifying a clause in the Electricity Act 2003 that charging for the purpose of charging an electric vehicle is classified as a service and that no licence is required for this business activity.

9.3.7 Other Initiatives

Income tax deduction of ₹1.5 lakh on the interest paid on loans to purchase electric vehicles. This amounts to a benefit of around Rs. 2.5 lakh over the loan period who would take loans to purchase electric vehicle. Similarly, GST on electric vehicles has been reduced from 12 to 5%. Similarly, GST on electric vehicles chargers has been reduced to 5% from 18%. Exemption of customs duties on certain parts of electric vehicles and capital goods is used for manufacturing of lithium ion cell. Exemption has been granted to all battery-operated transport vehicles and transport vehicles running on ethanol and methanol fuels from the requirement of permits. Ministry of Finance has rationalised the custom duty rates for all categories of vehicles—two-wheeler, three-wheeler, four-wheeler and buses and trucks which have been reduced to incentivize uptake of electric vehicles. This includes reduction of duty for completely knocked down and semi-knocked down unit of vehicles.

A pioneering initiative has been taken by Indian Space Research Organisation (ISRO) for commercialization of indigenously developed lithium ion battery technology. In this regard, ten firms have been shortlisted for the transfer of technology.

NITI Aayog has prepared a concessionaire Agreement for Public Private Partnership in Operation and Maintenance of Electric Buses in cities through Operating Expenditure (OPEX) Model. This would reduce the requirement of upfront capital as the lease would be signed on per km basis.

A grand challenge for developing the Indian Standards was initiated by Department of Science and Technology (DST), and Bureau of Indian Standards (BIS) has notified general requirements for electric vehicle charging based on CCS and Chademo charging standards.

Ministry of Housing and Urban Affairs (MoHUA) has prepared a draft for amendment of building code and town planning rules for provisioning for electric vehicle charging stations in private and commercial buildings. Ministry of Road Transport and Highways (MoRTH) has notified green number plates for battery-operated vehicles. Driving licenses are to be given for age group 16–18 years to drive gearless electric scooters/bikes up to 4 kWH battery size. Ministry of Road Transport and Highways has issued Draft Notification to exempt battery-operated vehicles from paying registration fees.

National common mobility card has been launched, which would allow payments across all segments including metro, bus, suburban railways, toll, parking, smart city and retail. These are bank issued cards on debit/credit/prepaid card product platform. The customer may use this single card for payments.

9.3.8 Electric Vehicle (EV) Policies Released by States and UTs

States and UTs have taken the lead in leapfrogging towards the electric mobility in the country. As on September, 2021, fourteen states and UTs have released the electric vehicle policy. Out of this, 13 states/UTs have finalised the EV policies, while six states have EV policies in the draft stage.

The 13 states/UTs which have finalised the EV policy are Andhra Pradesh, Karnataka, Kerala, Maharashtra, Tamil Nadu, Uttarakhand, Uttar Pradesh, Delhi, Madhya Pradesh, Telangana, Gujarat, Odisha and Assam.

The six states which have EV policies in the draft stage are Chandigarh, Bihar, Haryana, Himachal Pradesh, Punjab and Goa.

9.4 Challenges Which Need to Be Addressed to Boost Adoption of Electric Vehicles

NITI Aayog and World Energy Council [6] report mentions that the limiting factor of batteries on driving range could be addressed by developing ecosystem for fast charging or swapping of batteries. There is need to have proper infrastructure in every kilometer in dense areas. The report mentions that to make electric vehicles economically viable provide charging infrastructure and increase efficiency of vehicles. The other recommendations mentioned are policy for charging/swapping infrastructure; to make early impact focus on small and public vehicles; research and development; transforming auto-ancillaries: focus on power electronics industry and electricity distribution system/impact on the grid.

The issue of charging infrastructure, research & development, financing of electric vehicles, battery and cell manufacturing, proper regulatory framework, fiscal and non-fiscal incentives, availability of power and its infrastructure, consumer awareness, etc., need to be addressed to upscale electric vehicle adoption.

9.4.1 Increase Number of Charging Infrastructure in Country

There is need to increase the number of charging infrastructure as demand for electric vehicle increases in India. As on March 2021, India has 1800 charging stations. A study conducted by Singh et al. [22] for the Centre for Energy Finance shows that by 2030 around 29 lakh public charging stations are required to support EV adoption. Out of 29 lakh charges, around 21 lakh, i.e. 71%, would be low capacity charges used in two-wheelers and three-wheelers. The lack of space to charge electric vehicles is also a hurdle. Therefore, there is need to have dedicated parking space to charge electric vehicles.

9.4.2 Need to Push Research and Development (R&D)

Create a research & development fund (with private partnership) to be a leader in the next generation technologies. Create a dedicated R&D institution for EVs, comprising industry, academia and government. There is need to develop batteries which look beyond Lithium & Cobalt, and are made up of materials readily available in India. Strong R&D focus on fuel cells is also required.

9.4.3 Enable Financing for Electric Vehicles (EVs)

EVs are finding it very hard to access institutional finance. Even if they are able to get it, they are getting it at a high rate. India is being a cost-sensitive market, and upfront price of electric vehicles is being higher; without loans, the ecosystem will not take off. There is need to address the financing issue immediately.

9.4.4 Drive EV Manufacturing

There is need to take top high value components of EVs and aim to make India the export hub of these components. There is need to drive these on missions mode with specific states.

9.4.5 Drive EV Battery and Cell Manufacturing

India is importing batteries and does not have enough in-house supply of key raw materials, and this makes cell manufacturing a very costly affair. The manufacturers need to ensure adequate localization. The state governments in India are offering subsidies and addressing the infrastructure need for giga factories. There is need to sign Memorandum of Understanding (MoU) for procurement of raw material from countries rich in these resources.

9.4.6 Create Framework for Time of Day (ToD) Tariff

Discoms charge a flat rate of electricity throughout the day from users, even though the production is higher in the day (solar). Discoms need to have a lower price for electricity during day and higher during night. This will incentivize distributed storage where the users (including EV owners) buy electricity at a lower rate and may be use it or pump it back to the grid and earn money out it. The solar generation would go up, and cost of storage by discoms would go down as distributed storage (like batteries and EVs) would come up.

9.4.7 Consumer Awareness

Massive campaigns are required to influence consumer behaviour and to educate the citizens to adopt sustainable mobility choices. There is need to increase consumer awareness campaign for electric vehicle through print and social media.

9.4.8 Fiscal and Non-fiscal Incentives

There is need to provide fiscal and non-fiscal incentives by central and state government to upscale adoption of electric vehicles like subsidy, interest subvention, scrapping incentive, etc.

9.4.9 Congestion Pricing

EVs, though clean, do not solve the problem of congestion and do not incentivize people to move towards public transport. Do a pilot in a city, where the private vehicle owners are required to pay a small fee for entering in certain areas of the city. With the mandate of FAST tag across the country, the implementation would be seamless. The collected money can be put in a fund and used for promotion of electric public transport.

9.4.10 Mandate Battery Manufacturers to Procure Raw Materials (Say 15–20%) from Recycled Materials in India

A lot of e-waste is getting generated which is going into landfills without getting recycled. There is need to create a blockchain-based solution to track every battery from cradle to grave. The battery manufacturer's should be mandated to procure a certain percentage of raw material from the Indian recycle (& not import), which can be easily tracked by blockchain. This creates a market as well as demand for urban mining.

9.4.11 Centralised Authority to Spearhead the Electric Mobility Initiatives of Government of India

There is an urgent need for a centralised authority to spearhead the mobility initiatives of Government of India. Currently, there are various organisations engaged in different aspects of mobility at state and central level. A concentrated effort by a dedicated institution to coordinate with a wide variety of stakeholders, namely ministries, state governments, industry associations, manufacturers, technocrats, civil society, etc., is required.

India has immense potential to become leader in electric vehicles if challenges associated with adoption are addressed immediately. Electric vehicles are picking pace in India with increase in incentives, proper central and state government policies, with improving charging infrastructure and with announcement of manufacturing hubs and giga factories, with multiple industry players and start-ups entering EV market, with evolution around charging infrastructure and swappable batteries, etc. Government of India is making immense efforts to promote electric mobility.

References

1. IIT Kanpur. (2016). *Comprehensive Study on Air Pollution and Green House Gases (GHGs) in Delhi.*
2. IQ Air: https://www.iqair.com/world-most-polluted-countries. Accessed on August 2, 2021.
3. World Air Quality Report: Region & City PM2.5 Ranking (2020).
4. Invest India Website: https://www.investindia.gov.in/sector/automobile/electric-mobility. Accessed on September 6, 2021.
5. Automotive Mission Plan. A Curtain Raiser. (2016–26).
6. NITI Aayog and World Energy Council. (2018). *Zero Emission Vehicles (ZEVs): Towards a Policy Framework.*
7. Viswanathan, V., & Sripad, S. (2019). The key to an electric scooter revolution in India is getting the battery right. Quartz India (webpage): https://qz.com/india/1737200/an-electric-scooter-revolution-in-india-needs-better-batteries/

8. International Energy Agency (IEA). (2021). Global EV Outlook 2021. Accelerating ambitions despite the pandemic.
9. NITI Aayog and Rocky Mountain Institute. (January, 2021). Mobilising Finance for EVs in India: A Toolkit of Solutions to Mitigate Risks and Address Market Barriers.
10. NITI Aayog and Rocky Mountain Institute: India Leaps Ahead: Transformative mobility solutions for all. (2017). Link: https://www.rmi.org/insights/reports/transformative_mobility_solutions_india
11. Morgan Stanley. (2018). *The Next India, India's Transport Evolution.*
12. Bloomberg NEF: Battery Pack Prices Cited Below $100/kWh for the First Time in 2020, While Market Average Sits at $137/kWh. https://about.bnef.com/blog/battery-pack-prices-cited-below-100-kwh-for-the-first-time-in-2020-while-market-average-sits-at-137-kwh/. Accessed on 2nd August, 2021.
13. The International Council on Clean Transportation (ICCT). (2021). *FAME-II revisions spark hopes for a jump in electric two-wheeler sales in India.*
14. VahanSewa Dashboard by Ministry of Road Transport & Highways (MoRTH): https://vahan.parivahan.gov.in/vahan4dashboard/vahan/view/reportview.xhtml. Accessed on August 2, 2021.
15. Autocar India website: https://www.autocarindia.com/car-news/smev-cumulative-ev-sales-down-1941-percent-in-fy2021-420595. Accessed on August 21, 2021.
16. DHI. Ministry of Heavy Industries and Public Enterprises: National Electric Mobility Mission Plan 2020. https://dhi.nic.in/writereaddata/content/nemmp2020.pdf
17. Press Information Bureau (PIB): https://pib.gov.in/PressReleasePage.aspx?PRID=1576607. Accessed on August 2, 2021.
18. Present status of FAME II: https://fame2.heavyindustry.gov.in/. Accessed on September 6, 2021.
19. Invest India Website: https://www.investindia.gov.in/team-india-blogs/opportunities-ev-battery-and-cell-manufacturing-india. Accessed on August 8, 2021.
20. Press Information Bureau (PIB, 2019): Mobility Solutions get a boost, National Mission on Transformative Mobility and Battery Storage approved by Cabinet, Mission to promote Clean, Connected, Shared and Holistic Mobility Initiatives, Phased Manufacturing Programmes to be launched for batteries and electric vehicle components. Accessed on July 5, 2021. https://pib.gov.in/PressReleseDetail.aspx?PRID=1567807
21. Gazette Notification. (2021). Production Linked Incentive (PLI) Scheme, National Programme on Advanced Chemistry Cell (Acc) Battery Storage. Accessed on July 5, 2021. https://dhi.nic.in/writereaddata/UploadFile/ACC%20Scheme%20Notification%209June21.pdf
22. Singh, V. P., Chawla, K., & Jain, S. (2020). CEEW. Centre for Energy Finance. Financing India's Transition to Electric Vehicles. A USD 206 Billion Market Opportunity (FY21–FY30). (December, 2020).

Chapter 10
Sustainable Infrastructure Development in India: Drivers and Barriers

Muhammadriyaj Faniband, Kedar Vijay Marulkar, and Pravin Jadhav

Abstract India is one of the densely populated countries in the world and currently stands second in population next only to China. Therefore, the infrastructure requirement in the country is also considerably high. Unfortunately, the development in infrastructure has taken place, but it lacks proper policy and long-term vision. Infrastructure development must have a bearing on policy that would last long and would have consistency with vision of overall development plans of any government. Particularly, sustainability has not been considered while infrastructure development was going on. In this context, the chapter studies the sustainable infrastructure development in India. This chapter also focuses on drivers and barriers in sustainable infrastructure development. Slightly, India is now moving towards sustainable infrastructure developments. The recent schemes of the government give a clear hint to make a radical shift towards sustainability. The growing focus of the government is on the activities related to sustainable infrastructure.

Keywords Sustainable infrastructure · Infrastructure · Finance · Development · India

10.1 Introduction

India is set to become the third largest country in the world in terms of construction market. Massive budget allocations, huge subsidies along with tax concessions and government initiatives have made infrastructure sector a booming sector in Indian economy since last few years. Recently launched government schemes are adding

---

M. Faniband (✉)
Christ Academy Institute for Advanced Studies, Bengaluru, India

K. V. Marulkar
Department of Commerce and Management, Shivaji University, Kolhapur, India
e-mail: kvm_commerce@unishivaji.ac.in

P. Jadhav
Institute of Infrastructure, Technology, Research And Management (IITRAM), Ahmedabad, India

© The Author(s), under exclusive license to Springer Nature Singapore Pte Ltd. 2022
P. Jadhav and R. N. Choudhury (eds.), *Infrastructure Planning and Management in India*, Studies in Infrastructure and Control, https://doi.org/10.1007/978-981-16-8837-9_10

to the glory of glittering infrastructure market in India. While it focuses on development of economy as a whole, infrastructure development may also become cause of concern in the days to come if it ignores the environmental issues and sustainability. Hence, the time has come for us to think of sustainable infrastructure development instead just infrastructure development. Infrastructure development is looked as an important indicator in growth and development of economy. Through infrastructure, connectivity of various regions increases with each other, it facilitates improved supply chain, and it can lead to provision of clean water and electricity apart from basic social requirements like health and education. The infrastructure development essentially involves creation of new jobs which is another indicator of overall development of an economy. Thus, in general, infrastructure development facilitates growth of economy. The present paper is an attempt to discuss the issues arising in infrastructure development in a critical manner so as to avoid imbalance of development and also to throw light on sustainability which is becoming largely important aspect in every walk of life.

India is going to be benefitted from the plans of the government apart from the collaboration with various international groups which are coming forward for sustainable development and infrastructure associated with it. The banking system is also providing more and more assistance to the projects which are related to sustainable infrastructure. Further, most importantly the mindset of the people is also now changing. The people are shifting their traditional infrastructural requirements to innovative infrastructure projects involving sustainable development even if they cost more. Similarly, the UNO has also given the Sustainable Development Goals (SDG) which involve such sustainable infrastructure projects.

Sustainable infrastructure development is 'balancing between environment protection and economic prosperity for the long-term benefits. This refers to the principles of sustainable infrastructure development' [1]. 'Infrastructure development which values the environment can only lead to sustainable infrastructure development' [2] Thus, it can be summarized that sustainable infrastructure development is striking the balance between environmental and ecological protections vis-à-vis development which would benefit for future generations.

10.1.1 Prologue: Infrastructure Financing in India

Infrastructure financing in India in past was dominated by a few financial institutions. Industrial Finance Corporation of India (IFCI, 1948), Industrial Credit and Investment Corporation of India (ICICI, 1955), Industrial Development Bank of India (IDBI, 1964) started financing the infrastructural development in India in early independence period. State Industrial Development Banks (SIDBI) were also operating in various provinces. Thus, India has a long tradition in infrastructure financing; but during that period, sustainability was not given much importance. As economy needed boost and the Second Five-Year Plan focused on industrial development, lot of attention was given to create infrastructure and naturally, very less or negligible attention towards sustainability was given. Later, Rural Electrification Corporation (REC, 1970), Housing and Urban Development Corporation (HUDCO, 1970), National Bank for Agriculture and Rural Development (NABARD, 1981), Export–Import Bank of India (EXIM, 1981), Power Finance Corporation (1986), National Housing Bank (1988) started taking active part in infrastructure development in specific area. (For example, PFC devoted to only power sector, NABARD devoted to agricultural and rural development).

Kumar [3] stated that after inception of economic reforms in the year 1991, the working of National Development Banks (NDBs) is totally changed. Initially, NDBs namely IFCI, ICICI, and IDBI were dominant in the area of infrastructure finance, but gradually, they lost their support provided by the government as a part of reforms when the government left infrastructure development with private operators to a large extent. The real momentum in infrastructure finance was seen after introduction of economic reforms in India in the last decade of twentieth century. IDFC, IL and FS, L&T Infra Finance were few examples who started financing infrastructure development in a massive way. During the same time, the government (both central and state) started devising various schemes which promoted infrastructure financing in a big way. The question, however, was whether this boom was going to be sustainable! Recently, there has been lot of debate and discussion on sustainability particularly in terms of development along with environment protection. Unprecedented and at the same time, uncontrolled infrastructural development may be fruitful in the short term, but in the long run, it may have devastating effects on environment. Global warming, untimely rain, draughts, hurricanes are few disasters which we have been experiencing since recent past. Thus, time has come to think of sustainable infrastructure development financing instead of just GDP growth and industrial development.

10.1.2 Economic Reforms and Infrastructure Financing in India

After getting independence, India adopted mixed economy as a model of economic development. The Five-Year Plan model for economic development was initiated. The First Five-Year Plan was focused on rural development and basic amenities for independent India. In the second plan, however, the importance was given to industrial development. This paved way for infrastructural development in India. Roads, railways, air transport started gaining importance as a way of transportation of goods and services, and at the same time, the large projects like national highways, dams, education, and health infrastructure were also undertaken to have a holistic development. Though the space of economic and industrial development was low and infrastructure development was also not gaining that pace which was expected, it was consistent and waiting for some drastic things to happen. The major breakthrough was experienced in the year 1991 when infrastructure sector was made open. The development which required huge infrastructure were opened for private sector, and this move resulted in attracting bulk investment in infrastructural projects. This investment was from indigenous investors as well as from foreign investors. Foreign portfolio investors and foreign institutional investors started making investment in infrastructural projects in India. During this phase, sustainability was not a buzzword as it is today. But as the time progressed and industrial development gained momentum, sustainable infrastructure was also thought of. Development from IT, ITES drastically changed the face of many businesses. Infrastructure was also not exception to that. At the same time, due to the advent of IT, the focus was also given to sustainable models of infrastructure development, and accordingly, the policies of the government and local authorities also started promoting the projects involving sustainable infrastructure development.

Normally, investment in Indian private sector follow 70:30 proportion of debt and equity, respectively. Therefore, it was needed to promote financial institutions which would provide this intended debt funding to the infrastructural projects. It is in this context that new age National Development Banks (NDBs) namely IDFC and IIFCL were been created in 1997 and 2006, respectively. Both of these institutions mobilized the resources from the different sources which were unconventional and innovative at that time. It is expected that India may need around US$3.5 trillion in infrastructure investments by 2030 as estimated by the Boston Consulting Group. Innovative financial products are necessary apart from the traditional avenues to fulfil this requirement. Emergence of the BRICS Bank and Asian Infrastructure Investment Bank (AIIB) is an important development in infrastructure development and ultimately that would result in emergence of and shift towards sustainable infrastructure in the near future.

10.2 Drivers of Sustainable Infrastructure Development in India

Infrastructure Finance Companies (IFCs) were launched as a new category of Non-Banking Finance Companies from the year 2010 by the Reserve Bank of India. Keeping aside the traditional way of working, these IFCs started looking for some innovative sources to raise the finance.

10.2.1 Foreign Exchange Reserves for Infrastructure Development

In India, infrastructure development through IIFC Ltd., a UK-based financial institution was initiated as innovative weapon to apply foreign exchange reserves for infrastructure development. IIFC worked as a special purpose vehicle (SPV) to issue bonds in foreign currency. It also started as a co-financing Indian companies which were promoting infrastructure projects. Euro Commercial Bonds were issued for such projects which attracted international investment in infrastructure in India [4].

10.2.2 Green Bonds and Infrastructure Bonds

National Development Banks like IDFC started raising funds for infrastructure development in India. Some of the other companies like PFC, REC, IREDA were also raising cheaper, long-term resources for investment in energy conservation and generation. The benefits were given to the investors in terms of tax savings. These bonds, which were devoted for generating and conserving energy and also for promoting renewable energy, were named as Green Bonds [5].

10.2.3 Infrastructure Debt Funds (IDFs)

IDFs were launched to facilitate investment where local of foreign institutional investors, including insurance and pension funds were allowed to invest. This was specifically for refinancing existing debt of infrastructure companies. Thus, banks were given additional room to finance fresh infrastructure projects which were mostly in the form of Public–Private Partnership (PPP) model. IIFCL and IDFC were the initial players to raise the IDFs for infrastructure.

10.2.4 Intermediating Resources from Multilateral Development Banks (MDBs) and Bilateral Donors

International funding agencies like Asian Development Bank (ADB) and International Finance Corporation (IFC) and other private players started supporting sustainable infrastructure projects, especially since 2014. Initially, this was done by supporting solar projects. Later on, millions of dollars were provided for renewable energy projects through MDBs and such bilateral donors.

10.2.5 National Investment and Infrastructure Fund (NIIF)

The central government created NIIF after the union budget 2015–16 on the basis of some policy prescriptions contained in that budget. The aim of this fund was to promote infrastructure financing in India. It was established with a provision of Rs. 40,000 crores. The government intended a contribution of 49% in it and the remaining amount contributed from including PSUs, SWFs, pension funds, provident funds and National Small Savings Fund. NIIF is expected to liquidate public assets for infrastructure development. SEBI recognized this fund in December, 2015. As this involves generating funds from the public, NIIF has also been recognized [6].

10.2.6 National Clean Energy Fund (NCEF)

Sustainable infrastructure projects like solar power generation and other renewable energy projects are regarded as highly capital-intensive projects. Therefore, cost of capital is an important element in determining its viability. For this purpose, the NCEF was established. This agency has provided funds to especially renewable energy projects. The leading public sector bank SBI has also contributed a lot in renewable energy projects.

10.2.7 Ujwal DISCOM Assurance Yojana (UDAY)

UDAY is similar to refinance agency. It is not specifically providing finance to the infrastructure projects. But this scheme is designed to recast debt for state-run power distribution companies. The loans given by PFC and REC to these state-run companies are restructure through this scheme. Thus, PFC and REC can free their funds and mobilized the same for supporting renewable energy investments including through refinancing and take-out finance. This lending is expected to reach to the tune of a thousand billion rupees by the end of this decade [7].

10.2.8 India Infrastructure Project Development Fund (IIPDF)

Providing a channel or mechanism to finance bankable projects through competitive bidding process has been an important missing link in financing infrastructure projects. It was needed to have an advisory panel of experts who would determine the viability of the project as well as if needed, provide guidelines for the same. With this intention, IIPDF was established in the year 2007 with an initial contribution of Rs. 1000 million. Overall development of the project is monitored by IIPDF, and it also promotes PPP model to generate finance needed for infrastructural projects. IIPDF along with IL&FS started funding large infrastructure projects, like transport, ports, water and power infrastructure. IIPDF ensures all finance right from inception, progress and development of projects and also undertakes the downside risk up to financial closure [8].

10.3 Models of Infrastructure Financing

As a part of initiating a new model of infrastructure financing, IIFCL along with other National Development Banks started take-out finance as innovative model. Similarly, other credit enhancement schemes were also started for promoting infrastructure development finance.

10.3.1 Senior-Subordinated Debt

Commercially viable and high-valued projects were financed by IIFCL through consortium. Here IIFCL becomes senior agency which facilitates financing infrastructure projects involving large capital outlay.

10.3.2 Take-out Finance

In India, still commercial banks are leading in infrastructure financing. Due to the mechanism and working of commercial banks, there is a possibility of mismatch between assets and liabilities of these banks. Therefore, a refinancing authority or a third party like IIFCL takes over loans provided for infrastructure projects from the books of the banks. This facilitates them in making funds available for new infrastructure projects.

10.3.3 Credit Enhancement Scheme

Under credit enhancement scheme, the intermediary like IIFCL, ADB offers assurance to the credit to the bonds. These bonds are mainly issued by infrastructure companies. Subject to 50% of total project bond credit enhancement was given to the bonds with credit rating of AA.

10.3.4 Refinance Scheme

Specialized subsidiaries in sustainable finance were needed who could refinance the loan given by commercial banks of financial institution providing finance to such projects at concessional rate of interest. PFC, IIFCL have refinanced lot of schemes, especially in green power or renewable energy projects [9].

With the missions like housing for all or smart city campaign, there is an opportunity for making investment in building infrastructure to be erected in line with such schemes. The central government is providing large amount of funds for these dream projects. Considering nation-wide presence and participation for these schemes, huge finance is needed which will be made available by the government through leading banks or financial institutions. Development of airports was expected from the Regional Connectivity Scheme (RCS). Automatic route allowed even to the extent of 100% of FDI in such scheme. Projects worth Rs. 586.20 crore were sanctioned through the 'North Eastern Region for Social and Infrastructure Development Fund (SIDF)' [10]. This is going to play and important role in infrastructure development of the nation.

The government of India has launched various schemes and kept aside the budget for infrastructure financing. Of late, there has been increasing focus on sustainable infrastructure financing which can be seen from the following budget allocation from the Union budget 2020–21 [11]. The budget allocated to transport sector was Rs. 1.69 lakh crore, while Rs. 38,637.46 crore was allocated for the development of post and telecommunications departments. Similarly, capital expenditure for Indian railways was given boost by making provision of Rs. 72,216 crores. Rs. 91,823 crores were kept aside for road transport highway. A provision of Rs. 888 crores were made towards the upgradation of Government medical colleges, while Rs. 1361.00 crore were provided for government health institutions. Pradhan Mantri Gram Sadak Yojana (PMGSY) was given a fund of Rs. 80,250 crores, and Bharatmala phase 2 to be implemented with a provision of Rs. 8000 crores. All these schemes are long-term schemes, and they are meant for development of sustainable infrastructure in India.

10.4 Barriers of Sustainable Infrastructure

Infrastructure development without due care can be disastrous. Considering the environment, economy and the society, development should be thought of. There are few barriers in development of sustainable infrastructure.

10.4.1 Large Capital Outlay

Sustainable infrastructure projects are not established projects. They are in nascent stage, and to develop such infrastructure, huge capital is required. Therefore, even big corporate houses also are unable to easily execute such projects. Therefore, there is large dependence on government to develop sustainable infrastructure.

10.4.2 Technology

The capital-intensive sustainable projects also require modern and innovative technology. The cost of such technology is also not affordable, and apart from this, the availability of this technology is also a matter of concern. High-end technology projects are beneficial in long term. But to start with, the investment required in these projects may not be affordable to those who are interested to launch such projects.

10.4.3 Planning

Most of the projects look attractive on paper. But when the time comes for actual implementation of such projects, there are some bottlenecks. Proper planning will help to sort out these issues. Initially, the project is started without proper planning and only with enthusiasm. Therefore, it may face problems in future. With the help of proper planning, consideration of aspects like budgets, cash flow, coordination, execution, probable hindrances can be identified in advance and one can thing of overcoming these problems.

10.4.4 Mechanism/Execution

The government comes out with innovative projects which are sustainable. The people/agencies interested in implementing these projects prepare the project reports and submit it to the concerned agencies. But when these projects are selected or implemented, there needs to be proper mechanism to monitor the progress as well as determine the viability of the project. Due to non-availability of such mechanism, there is a possibility of red tapism which is major hurdle in successful implementation of such projects. There needs to have a mechanism which controls and coordinates the project right from its inception to the conclusion so that effective implementation is possible.

10.4.5 Negligence Towards Socially Relevant Projects

Most of the infrastructure development projects are concentrated on urban and metropolitan areas. Even in urban areas, basic amenities are still not available to all the people. The basic need of the people is education and health. In urban as well as rural areas, there is a lack of infrastructure in relation to education and health. Private players are providing these facilities to the people. But these facilities are unaffordable to the common people. Hence, there is a scope in developing infrastructure required for education and health even at rural areas and even in urban areas also. These projects will be viable as they will be related to the essential requirements of large strata of the people. In the long run, the negligence towards basic facilities and their infrastructure may create adverse situation.

10.5 Strategies for Sustainable Infrastructure Finance

For policy makers and government bodies, it is the high time to give importance to the infrastructure projects which are sustainable. Gone are the days when only financially viable projects were held important. Now it is necessary to support projects which are less harmful to the environment.

Provision of basic infrastructure, through PPP model, is the fast-growing model of development which has been proved in the recent past. Apart from physical infrastructure projects which are concentrating in socially relevant projects like health and education, also need attention. However, if basic facilities are not present these programmes would have no meaning. Poor people need to bring in the development, and schemes for them should not be 'poor'. Poor and downtrodden people given a chance can also contribute to the economy. This strength needs to be addressed for quality solutions in implementing substandard measures launched through the 'poverty' mindset. This requires proper thinking and vision. If these sections are considered, then only one can assure sustainable development. Plans have to be worked at operational level, understanding their needs and technical viability, so as to avoid problems at later stages. Revision in such plans needs to made in consultation with all the concerned stakeholders.

Eight core infrastructure industries are the major backbone for infrastructure development. Nowadays, with increasing importance being given to sustainability in infrastructure, these core industries are also spending a lot on sustainability while keeping an eye on economic development.

10.6 Epilogue: The Road Ahead

At present, commercial banks, National Development Banks, refinance agencies are providing finance to the sustainable infrastructure. Apart from this, investments are also attracted through a variety of ways for infrastructure development. The funds are now being mobilized from diverse domestic and international sources. Sustainable infrastructure needs long-term resources. The project involving sustainable infrastructure need to be monitored and controlled properly. A mechanism needs to be devised to monitor their progress. Further, there needs to be coordination between various financing and refinancing agencies so that required focus can be given for inclusive development. Otherwise, the financing may get concentrated in the hands few big corporates. The provision of finance for socially relevant projects including health and primary education to the rural and urban areas needs to be made available effectively. For monitoring the progress and intermittent reviews of the ongoing projects, there needs to be some agency which would review the project periodically and suggest/recommend necessary actions if needed. The refinancing agencies are also doing their job of providing finance to the sustainable infrastructure projects. Takeout finance and credit enhancement strategies help in freeing the balance sheet of banks from the evils. Still lot of burden of infrastructure development rests on commercial and nationalized banks in spite of specialized agencies being present in the sector. Thus, now the time has come to seriously think over and determine the priorities. According to the priorities and goals, we need to march ahead. UNO has also given 17 goals of sustainable development (SDGs). Throughout the world, sustainability is being thought of and being executed. But there is a long way to go as we have started to feel the heat of global warming, cyclones, untimely rains, draughts, etc. In order to overcome this, one country can fight alone, but it is equally important that through agencies like UNO or other international groups, the world comes together to build sustainable infrastructure and various countries need to join their hands together, keeping aside their political and financial goals for making earth a better place to live.

References

1. Chan, I. Y. S., & Liu, A. M. M. (2012). Antecedents of innovation climate in construction firms in Hong Kong. *International Journal of Construction Management, 12*(4), 37–46. https://doi.org/10.1080/15623599.2012.10773199
2. Stoddart, H. (2011). A Pocket Guide to Sustainable Development Governance. Commonwealth Secretariat Stakeholder Forum.
3. Kumar, N. (2016). National Development Banks and Sustainable Infrastructure in South Asia. GEGI Working Papers Series.
4. Reserve Bank of India. (2010, August 17). *Infrastructure Financing: Global Pattern and the Indian Experience: RBI Staff Study*. Retrieved from https://rbi.org.in/Scripts/BS_PressReleaseDisplay.aspx?prid=23001
5. Bhaskar, U. (2015, June 03). REC, PFC, IREDA, others to raise Rs 5000 crores via tax-free bonds. *Live Mint*.

6. Sikarwar, D. (2015, December 30). Modi Sarkar's Rs. 40Kcr Infra Fund hits Road. *The Economic Times.*
7. Singh, S. (2016, January 12). PFC, REC to lend Rs 1 L Cr to renewable energy sector, *The Economic Times.*
8. Government of India. (n.d.). *Scheme and Guidelines for India Infrastructure Project Development Fund.* Retrieved from http://mohua.gov.in/upload/uploadfiles/files/Guideline_Scheme_IIPDF.pdf
9. Power Finance Corporation. (2014). *Annual Report 2013–14.* Retrieved from https://www.pfcindia.com/DocumentRepository/ckfinder/files/ann_rpt_2013_14.pdf
10. India Brand Equity Foundation. (2021, January). *Indian Infrastructure Sector in India Industry Report.* Retrieved from https://www.ibef.org/industry/infrastructure-sector-india.aspx
11. Government of India. (2020). *Union Budget: 2020–21.* Retrieved from https://www.indiabudget.gov.in/budget2020-21/

Part IV
Developing Digital Infrastructure

Chapter 11
Recent Trends in Digital Infrastructure in India

Krishna Teja Perannagari and Vineet Gupta

Abstract In the modern era, digital infrastructure plays a vital role in the economic development of a country and is recognised as a key contribution factor for fourth industrial development. Despite the increasing dependence of economic activities on digital infrastructure, emerging economies are struggling in their efforts to build a robust digital infrastructure due to weak affordability index and lack of coordination between various stakeholders. The current chapter discusses India's digital transformation, highlighting the various programs and initiatives taken by the Indian government to foster the goal of Digital India. The authors also shed light on recent technological trends influencing the development of digital infrastructure and suggest measures to encourage the adoption of latest technology and promote investments in digital infrastructure projects.

Keywords Digital infrastructure · Digital India · India Stack · New educational policy

11.1 Introduction

In the twentieth century, the primary focus of different nations was on developing physical infrastructure such as roads, irrigation canals and power stations, as they supported human flourishing by enabling trade and improving human productivity. However, in the twenty-first century, because of changing requirements of individuals and businesses, nations have shifted their priority towards developing infrastructure that supports the provision of technology-based services. The term *digital infrastructure* refers to a wide range of technologies that help in delivering services over the Internet [1]. Examples of digital infrastructure include broadband network, mobile

---

K. T. Perannagari
MICA - The School of Ideas, Ahmedabad 380058, India

V. Gupta (✉)
School of Business Studies, Sharda University, Noida 201310, India
e-mail: vineet.gupta1@sharda.ac.in

© The Author(s), under exclusive license to Springer Nature Singapore Pte Ltd. 2022
P. Jadhav and R. N. Choudhury (eds.), *Infrastructure Planning and Management in India*,
Studies in Infrastructure and Control, https://doi.org/10.1007/978-981-16-8837-9_11

services, data centres, communication equipment, cloud platforms, software applications, user devices, Internet-of-Things (IoT) devices and application programming interface (API).

Digital infrastructure is gradually becoming the primary driver of the country's economic development and exhibits a significant impact on the progress of the nation. The primary advantage of developing digital infrastructure is that it simplifies the process of automation and helps in reducing resource wastage. Realising the importance of digital infrastructure, many developed economies in Organization for Economic Co-operation and Development have built digital infrastructure to cope with high usage of Internet, and this substantially lowered the capital expenditure and trade costs [2]. However, a stable backbone for digital infrastructure is still lacking in many emerging economies, resulting in *Digital Divide* between developed and emerging economies. To overcome the digital divide, emerging economies need to increase their spending on digital infrastructure facilities and develop new business policies that would promote private and foreign investment in building digital infrastructure.

Many challenges exist in developing economies that hinder the provision of digital services to individuals, businesses and households. For example, in many rural regions and remote locations, the population is scattered across vast geography, increasing the cost involved in providing Internet services. Similar other barriers that are not specific to a particular geography include spending capacity, awareness about digital platforms and lack of technical skills required for incorporating technology-based services in one's personal/professional domain. To overcome these barriers, there is a need for coordinated efforts among different stakeholders, and developing countries should make use of rapidly emerging technological advancements such as 5G, machine learning, virtual/augmented reality, blockchain and IoT. In addition, countries should also focus on building a reliable and scalable digital infrastructure to prepare for the future by collaborating with various public and private sector entities in developed countries.

Before proceeding with the task of developing digital infrastructure for a country, many factors need to be considered, such as geographical distribution, purchasing power and infrastructure requirements. Replicating the template used in developed economies to build digital infrastructure for emerging economies may not be feasible in the long run, as each country has its own set of requirements and barriers. Similarly, global changes such as development of 5G technology, increasing cost of electronic hardware equipment and the emergence of pandemic also influence the digital transformation process of a country. For example, the COVID-19 pandemic accelerated the development of digital infrastructure in various African and Latin America countries as the dependency of people on Internet and smart phones increased tremendously. The current chapter describes the digital transformation process of Indian subcontinent and summarises the recent developments regarding digital infrastructure in India.

11.2 Digital Infrastructure in India

In the last decade, India has witnessed a digital transformation with government playing a central role and emerged as the second-fastest digitised economy next to China [3]. To foster India's digital revolution, Government of India has launched various initiatives such as Make in India, Digital India, Innovation Fund, Skill India and Start-up India to enable the delivery of various services through Internet. Digital India program launched in 2015 aims to make digital infrastructure as a basic utility for every Indian citizen by connecting even the remote areas with Internet technology. Figure 11.1 provides an overview of key focus areas of Digital India program. Another bold effort taken by the government was to be demonetise the existing high denomination bank notes and try to achieve financial inclusion by making Aadhar a platform for cashless transactions [4]. To further accelerate the development of digital economy, the government promoted uniform payment interface (UPI) and launched Jan Dan scheme to provide zero balance bank accounts.

Another key enabling factor for the success of digital initiatives is the presence of a large number of young citizens who are quick adopters of technology [6]. The Digital India program and other initiatives proved fruitful during the outbreak of COVID-19 pandemic as individuals were equipped with the ecosystem that would enable them to operate from their own homes using Internet facilities. Because of various initiatives from government and private sector, the number of Internet connections increased by 47 million between 2020 and 2021 Internet penetration rate to that 45% with 624 million connections [7]. Similarly, the number of mobile connections stood at 1.1 billion and exhibited a penetration rate of 79%. Statistics also indicate that the number of mobile connections increased by 23 million and the growth rate of mobile connections was 2.1% between January 2020 and January 2021 [8].

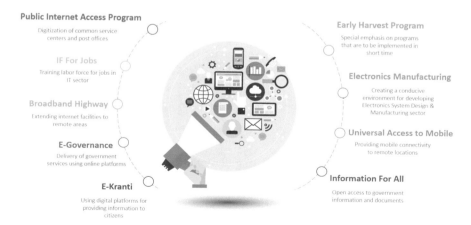

Fig. 11.1 Key focus areas of Digital India [5]

India also has the cheapest mobile data plans, with 1 GB of mobile data costing around $0.26, which is far less than the global average of $8.53, and average mobile data consumption stands around 200 MB per day [9]. Digital transactions are also witnessing a rapid growth in India, with UPI transactions occupying a majority stake in percentage of total cashless transactions. Recent estimates reveal that between December 2019 and December 2020 the number of banks offering services using UPI platform increased from 143 to 207 and value of UPI transactions rose from 2.20 lakh crores to 4.16 lakh crores showing a 105% growth rate [10].

COVID-19 pandemic also provided a new beginning for adoption of new technology for digital infrastructure. Global lockdown and quarantine measures increased people's dependency on Internet for performing even basic tasks. Big data analytics and machine learning were seen as the new tools to combat the pandemic. Similarly, work from home, online shopping, telemedicine and e-learning became the new normal stressing the need for high-speed Internet facilities. Due to a sudden spike in demand for videoconferencing, top multinational companies such as Google (Google Meet), Microsoft (Microsoft Teams), Cisco (Cisco WebEx) and Facebook (Messenger desktop) started providing their own video calling applications. Similarly, Indian start-up companies such as Inscripts (Namaste), 10Times (Floor app) and KnowledgeLens (KL Meet) started offering their own videoconferencing application to cater the growing demand for videoconferencing [11].

Similarly, the largest vaccination drive in the world took place in India, and CoWIN was developed as an extension of electronic vaccine intelligence network to implement and monitor the vaccination program in India. The rising digital footprint is also creating a new set of challenges that need to be addressed, such as the impact of censorship, privacy concerns and rising cyber-crime. In addition, a lot of progress is still to be done regarding the usage of data generated from the digital ecosystem and development of hardware manufacturing capabilities to promote digital equipment produced in India. To overcome these challenges and assist the integration of various entities in the digital infrastructure, technology frameworks and regulatory policies have emerged in recent years. The next section provides a brief overview of the India Stack architecture and explains its significance for developing digital solutions that enable paperless and cashless service delivery.

11.3 India Stack

India Stack makes use of India's existing digital infrastructure comprising various e-governance services such as Aadhaar, eKYC, eSign, DigiLocker and UPI to develop a software ecosystem that brings millions of Indians into formal economy. To develop the software ecosystem, India Stack provides four different technology layers comprising presence-less layer, paperless layer, cashless layer and consent layer. Figure 11.2 provides an overview of India Stack architecture. Aadhaar, eSign, DigiLocker and UPI are integrated into various layers of India Stack to offer cashless and paperless services. Aadhar is the key component of presence-less layer and

11 Recent Trends in Digital Infrastructure in India

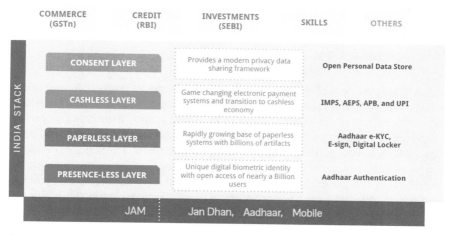

Fig. 11.2 Various components of Indian stack architecture [12]

helps the software application to identify the user based on the biometric details and 12-digit Aadhar unique identification number.

The paperless layer comprises digital locker and digital signature components. The services required for digital locker component are provided by DigiLocker by enabling the individuals to store and share important documents such as land records, college transcripts, address proof, driving license and medical records. The documents in DigiLocker also contain a digital signature of the issuing party and are deemed on par with original physical documents. eSign service provided by Controller of Certifying Authorities and eKYC services provided by UIDAI comprises the digital signature component and enables the individuals to sign a contract or document using biometric details and Aadhar authentication.

The third layer is composed of various cashless transaction services provided by government and banking institutions such as UPI, immediate payment service and Aadhar payment. The transaction is enabled by simple APIs that sit at the top of complex technical architecture and provides instantaneous payment facility between the sender and receiver. Finally, the consent layer is composed of third-party applications, such as EazyGov, Krishiyog, Gradufund, CancerRegis, and Journee. These applications follow the specification prescribed in India's Open API policy and use identity verified time bound consent tokens to share information with other applications in the India Stack ecosystem [13]. The next section describes various industries that are experiencing a rapid phase of digital transformation and explains the impact of COVID-19 pandemic on digital infrastructure development.

11.4 Industry's Leading Digital Transformation in India

11.4.1 Health Care

Indian healthcare system is characterised by the presence of both public and private entities with most of the public health systems focusing on providing primary healthcare services. Private hospitals in India offer a wide range of services and provide a good value for money when compared with hospitals located in developed countries, making India a preferred location for medical tourism. Despite the prevalence of a vast healthcare network, a wide disparity occurs in terms of quality, cost and coverage based on the geography in which the hospital is located. Table 11.1 provides an overview of Indian health sector. In recent years, digital technology is picking up its phase, with digital health records being one of the most widely used applications in healthcare industry [14]. To create digital health ecosystem and build public health infrastructure, Indian ministry of health and family welfare has launched a National Digital Health Mission that is used by over 12 lakh Indian citizens [15]. The main purpose of this mission is to make the health data accessible to various stakeholders involved in providing healthcare services without compromising on the privacy of health-related personal information.

Healthcare systems experienced maximum load during COVID-19 pandemic highlighting the role played by digitisation in building a robust public health infrastructure and ensuring health equity to vulnerable sections of the community. To deal with the current pandemic, most of the countries including India ramped up the production of medical equipment and developed new medical facilities by converting public and private spaces into COVID-19 care facilities. Despite several measures undertaken by the government to handle the COVID-19 situation, the imbalance between the demand and availability of healthcare facilities is growing because of the current surge in cases and emergence of new COVID-19 strains. Preliminary estimates reveal that availability of trained medical professionals would be the next major challenge for dealing with the COVID-19 cases as the pandemic is also exhibiting a serious impact on the health of the medical professionals [17, 18].

Table 11.1 Statistics related to Indian health sector [16]

	2016	2017	2018	2019	2020	2021	2022 (F)
Market size (US$ billions)	140	160	–	–	280	–	372
Health insurance premium (US$ billions)	3.82	4.65	5.9	6.58	7.04	8	–
Healthcare expenditure (% of GDP)	1.3	1.4	1.4	1.5	1.8	–	2.5
Doctors	–	–	1,154,686	–	1,255,786	–	–
Medical college	412	462	476	529	541	562	–
GDP per capita (US$)	1761	2014	2036	2181	2378	2578	2791

Similarly, there is also a need to focus on the medical cases related to non-COVID emergencies as patients suffering from other chronic illness are being ignored because of the diversion of medical professionals towards treatment of COVID-19-related cases. Boosting the technological infrastructure of medical facilities is deemed as the most effective solution for dealing with the shortage of trained medical personnel [19, 20]. Digital platforms that facilitate tele-consultation and videoconferencing are used by major healthcare providers in the private sector to treat the patients suffering from mild illnesses and to protect the healthcare professionals from COVID-19 pandemic. The primary challenge in offering tele-consulting services is the availability of Internet connectivity and proper institutional training for medical professionals involved in the tele-consultation process. To overcome these barriers, healthcare providers are focussing on providing training to impart necessary skills and knowledge required for efficient delivery of telemedicine services.

11.4.2 Education

India has the second largest workforce in the world, and education sector plays a very important role for development of necessary skills and knowledge, making the workforce employed in various sectors of Indian economy. Owing to the large population scattered across different geography areas, Indian education sector is characterised by different types of educational institutions that offer a wide range of schooling options including provision for vocational courses and distance education. Realising the importance of education for boosting the literacy rate and improving the wellbeing of citizens, the Government of India has introduced the right to education act in 2009, making education a fundamental right for every child in India [21]. To further transform the learning ecosystem, the Government of India has launched a new education policy in 2021. Figure 11.3 provides an overview of various reforms introduced in the new education policy. In the current era, digital infrastructure and physical infrastructure are considered two important faces for education delivery. In comparison with physical infrastructure, development of digital infrastructure requires fewer resources and enables access to interactive multimedia-based education content, allowing students to learn anywhere and at any time. Building digital infrastructure for existing educational institutions is further simplified by learning management solutions provided by Indian EdTech companies such as Classplus, Ken42 and Next education.

IT-enabled learning is also seen as a key differentiating factor for improving the learning outcomes and creating an experiential learning environment in schools and colleges. Realising the potential of digital education, many technology platforms such as Coursera, Udemy, Skillshare and Udacity have also started offering online courses on various subjects by partnering with universities, educational institutions and multinational companies. Despite the availability of various technologies that assist the development of digital infrastructure in the educational institutions, the adoption of digital learning is restricted primarily to metropolitan cities and urbanised

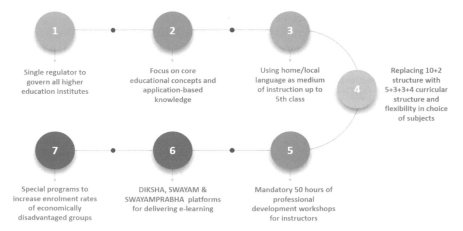

Fig. 11.3 Reforms in new education policy 2021 [22]

areas in India. Digital divide is one of the key contributing factors that has hindered the efficiency of digital learning initiative in the underdeveloped regions of India. The negative impact of digital divide on educational outcomes is further strengthened by the COVID-19 pandemic, as educational institutions are forced to conduct classes only on online mode, and most of the households in India lack access to computer and broadband Internet [23, 24].

Despite the increase in smartphone adoption rates and availability of cheaper mobile Internet, a significant number of households belonging to the socially disadvantaged class still lack basic digital infrastructure required for accessing online education, and there is a need to improve digital literacy by developing low-cost equipment that can be used for educational purposes and making Internet services available even in remote locations. To assist online education during COVID-19, Indian ministry of education has undertaken several steps including increased emphasis in the New Education Policy for building digital infrastructure and use of TV channels, radio and podcasts to deliver teaching material to students who do not have access to smartphones and Internet connectivity. Similarly, Digital Infrastructure for Knowledge Sharing (DIKSHA) online platform and SwayamPrabha TV Channels are developed extensively under the PM eVIDYA initiative to support online education in India [25].

11.4.3 Transportation

Transportation ecosystem is considered the backbone of a country's economy as it acts as an enabler for various economic activities happening in the country. Transportation infrastructure in India is still in the initial stages of development and is characterised by obsolete assets that need to be upgraded to save costs. Another major shortcoming of public transportation sector in India is a lack of coordination and heavy dependence on manual labour for repetitive tasks. Inefficiencies in existing transportation infrastructure exhibit a profound influence on economy and well-being of the individuals by extending the time taken for transit. Incorporating digital solutions into transportation ecosystem can lead to significant savings in cost by improving the efficiency of existing transportation facilities and helps in reducing dependency on personal transport, reducing the carbon footprint from vehicles [26]. Digitisation is often seen as a key for development of advanced transportation facilities such as high-speed rail networks and enables the usage of unmanned aerial vehicles or drones for transportation [27, 28].

To overcome the disadvantages associated with internal combustion engines running on fossil fuel, most of the countries including India are transitioning towards electric vehicles as they are more energy efficient and quieter. Another advantage of electric vehicles is that they are cheaper to maintain and provide huge cost savings due to lower running costs. Figure 11.4 provides the forecasts for electric vehicle penetration in India. Digitisation will also assist the government initiative to reduce dependency on crude oil imported from other countries by accelerating the phase of electric vehicle adoption and helps in providing AI-assisted smart driving features in electric vehicles [29]. Most of the developments related to digital infrastructure in transportation sector are concentrated in commercial logistics, and COVID-19 pandemic had a positive impact in this process because of the increased adoption rates of e-commerce industry. Indian government has also undertaken few important measures to digitise public transportation such as introducing FASTag to collect toll charges from national highways and using intelligent transport system in public network that provides various services such as global positioning system to track public transport vehicles, Internet ticketing system and enabling cashless digital payments.

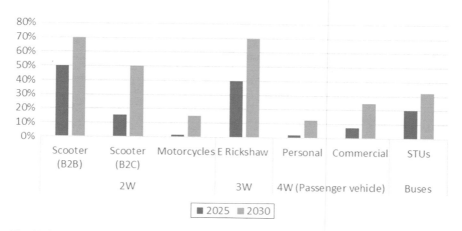

Fig. 11.4 Forecasts for electric vehicle penetration in India [30]

The major challenge associated with the digitisation process in public transportation networks is the lack of digital literacy among employees working in the public transportation sector, and this reduces the efficiency of the developed digital solutions. Similarly, because of the increase in complexity of the digital transportation network, maintaining and upgrading the systems requires specialised skills, and this will have a significant impact on the overall cost incurred in managing the public transportation infrastructure [31]. To justify the enormous costs involved in building such infrastructure, transportation projects need to achieve sufficient economies of scale, and reaching this milestone is another challenge faced by public transportation sector. To overcome the challenges related to the development of transport infrastructure, public transport corporations are employing public-private partnership model. Similarly, to compensate for the lack of digital skills among employees, public transport corporations are administering specialised training programs and certification courses [32].

11.4.4 Financial Services

Financial services include a wide range of businesses such as banking, insurance, mutual funds, taxation, auditing, accounting and brokerages that are primarily concerned with money, wealth and investment management. Table 11.2 provides an overview of Indian financial sector. Digitisation in financial services is one of the primary targets of Indian government, and a lot of schemes are put in place since the early 90s with the first step being the use of electronic fund transfers and automated teller machines. To promote cashless transactions, National Payments Corporation of India and Reserve Bank of India have launched a UPI system and integrated it with services provided by Indian banks and digital payment gateways such as Google Pay, PhonePe, Paytm and BHIM UPI. Similarly, digitisation has also witnessed a high success rate in mutual funds and stock market transactions, with most stockbrokers and mutual fund companies offering a digital platform for their customers.

The COVID-19 pandemic has increased the pace of digitisation of financial services in India with most taxation and auditing service providers adopting digital technology to provide the services amidst the COVID-19 lockdowns. Similarly, digital banks that operate on paperless banking system such as Airtel payments bank and Paytm payments bank are also launching in India. Introduction of digital services and financial sector also opens certain new challenges among which cybersecurity and cyber-crime are the most prominent one [34]. To combat most of the Indian banks and digital service providers are constantly upgrading their systems and providing necessary cybersecurity training to the employees. Similarly, Reserve Bank of India (RBI) and Government of India have also taken proactive steps to educate the customers about various scams and phishing frauds happening in the country. Another recent development in digitisation of financial services is the increase in transactions related to cryptocurrencies, such as bitcoin, Ethereum, Tether and Binance Coin. To make use of the advantages provided by cryptocurrencies, the Indian government

11 Recent Trends in Digital Infrastructure in India

Table 11.2 Indian financial sector performance between 2016 and 2021 [33]

	2016	2017	2018	2019	2020	2021 (F)
Mutual fund assets (US$ billions)	252.06	272.62	331.42	340.48	404.73	438.27
Companies listed on NSE and BSE	7719	7651	7501	7586	7172	–
Amount raised by IPOs (US$ billion)	2.31	4.54	13.01	2.85	2.87	–
Number of high-net-worth individuals in India	200,000	255,000	256,000	263,000	–	–
Life insurance premium—new subscriptions (US$ billion)	21.5	27.2	30.1	30.7	37	–
Life insurance premium–renewals (US$ billion)	35.3	37.7	41	42	30.6	–
Non-life insurance premiums (US$ billion)	14.72	19.11	23.38	24.32	26.86	22.61
Non-banking financial companies' public funds	278.23	332.46	407.11	470.74	–	–
Gross national savings (% of GDP)	30.4	30.2	29.6	28	–	–
Turnover for derivatives segment (US$ trillion)	9.9	14.07	25.6	33.99	49	72.32

and RBI are considering the options for launching a digital currency which is backed by the RBI [35].

11.4.5 Retail

Indian retail sector is one of the fastest-growing sectors in the world and is witnessing a fresh wave of digitisation as COVID-19 pandemic resulted in a shift towards online shopping because of the global lockdowns. Online shopping is still in the nascent stages in India as local mom-and-pop stores are preferred by the Indian consumers because of the convenience and localisation in the shopping process. Because of social distancing norms and fear of infection, majority of Indian consumers have now transitioned towards online shopping, and this is driving the growth of e-commerce players such as Amazon and Flipkart. Grocery and apparel/fashion product categories are driving this growth, and multinational companies such as Tata and Reliance are also entering this segment to capture the growing market [36]. Figure 11.5 provides an overview of most valuable product segments in Indian e-commerce sector. The transition towards online shopping and the rise of e-commerce players has sped up

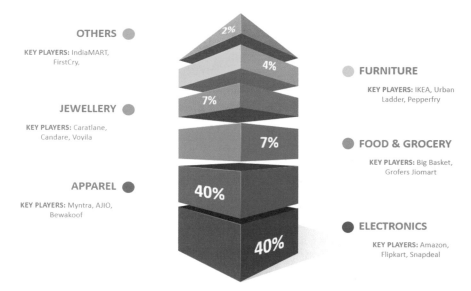

Fig. 11.5 Top segments in retail e-commerce by value [37]

the phase of digitisation, and local retail providers are also digitising the services by partnering with software application providers.

Despite the growing demand for e-commerce and initiation of programs like Digital India, Make in India and Skill India, the penetration of digital infrastructure that can support online retail services is still low, and there is a need to improve cooperation between corporate players and Indian government. Besides digital infrastructure, physical infrastructure is also underdeveloped, and there is a scope for corporate entities to collaborate with government bodies and work towards developing the infrastructure facilities required for capturing the untapped markets in sparsely populated areas. Recognising the importance of digital infrastructure, the Government of India has announced digitisation of retail as one of the key focus areas in the new retail policy proposed in 2021 Indian budget [38]. Multinational companies in the private sector are also investing in digitising retail sector by developing new application services for managing retail operations such as inventory management, ledger management and supply chain management. Google has also launched a fund called *Google for India Digitisation Fund* focusing primarily on digitising the services used by retail players [39]. Omni channel retailing that integrates online and off-line markets is often seen as the key for building the retail ecosystem in India. To assist in this process, reliance retail is developing its own JioMart platform and trying to build a full stack data-driven solution for small retailers [40].

11.5 Future Trends in Digital Infrastructure

COVID-19 accelerated the phase of digital technology adoption, making digital infrastructure vital component for functioning of almost every aspect of modern society. The need for Internet facilities and computing devices will increase tremendously in the upcoming future with an increase in usage of digital services to perform routine tasks. To make the ever-growing need for high-speed Internet connectivity, countries have started deploying 5G technology and are using fibre-optic connections to deliver reliable services directly to the end-user. Similarly, organizations are upgrading their existing standalone computing systems to work on cloud-native ecosystem as it helps in improving the performance of existing applications. The shift towards cloud-infrastructure also helps the organizations in making their computing system compliant with modern data protection standards, such as General Data Protection Regulation and California Consumer Privacy Act [41].

Rise of digital services also contributes to greenhouse emission as the power required for operating data centres that store and process large amount of information is huge, and a large portion of this power is still generated from fossil fuel in emerging economies. To reduce the environmental impact of digital transformation, countries are employing edge first paradigm, shifting the data close to the originating source and reducing round-trip latencies. Data centres are also trying to become carbon neutral by shifting towards renewable sources of energy and using coolant technologies that minimise electricity consumption. Data generated from a wide range of digital applications also improve the efficiency of existing machine learning and artificial intelligence algorithms, resulting in the efficiency doubling every 16 months [42]. Emerging economies like India can benefit a lot from the recent trends in digital infrastructure and require digital leaders who can prepare India to embrace these changes and position India as a technologically advanced nation.

References

1. Henfridsson, O., & Bygstad, B. (2013). *The generative mechanisms of digital infrastructure evolution.* MIS Q 907–931.
2. Cassiman, B., Sieber, S. (2007) The impact of the internet on market structure. *Handbook of Information Technology in Organisation and Electronic Mark*, 299–322.
3. Gill, P. (2019). India is digitising fast but the pace is leaving many behind. *Bus. Insid.*
4. Kumar, A. (2017). Demonetization and cashless banking transactions in India. *International Journal of New Innovations in Engineering and Technology, 7*, 30–36.
5. DigitalIndia.gov.in. (2015). *How Digital India will be Realized: Pillars of Digital India.*
6. Silver, L. (2019). In emerging economies, smartphone adoption has grown more quickly among younger generations. Pew Research Center.
7. Keelery, S. (2021). Internet penetration rate in India 2007–2021. *Statista.*
8. Kemp, S. (2021). DIGITAL 2021: INDIA. *Data Reportal.*
9. Roy, P. (2019). Mobile data: Why India has the world's cheapest. *BBC.*
10. Soni, S. (2021). UPI ends 2020 on high note, scales past Rs. 4-lakh-cr milestone in December; volume up 70% from year-ago. *Financ. Express.*

11. Gopal, S. (2020). Zoom alternatives: Here are the Indian startups taking on the popular video conferencing app. *Indian Express*.
12. IndiaStack.org. (2016). *India Stack: Towards presence-less, paperless, and cashless service delivery*.
13. Bhalla, T. (2017). *12 startups from #BuildonIndiaStack venture pitch that are leveraging IndiaStack*. Your Story.
14. Mathur, N. (2019). Digital health technology can revolutionise healthcare in India. *Mint*.
15. Sharma, N. C. (2021b). National digital health mission sees generation of 12 lakh IDs. *Mint*.
16. IBEF.org. (2021a). *Indian Healthcare Industry Analysis*.
17. Nagesh, S., & Chakraborty, S. (2020). Saving the frontline health workforce amidst the COVID-19 crisis: Challenges and recommendations. *Journal of Global Health, 10*.
18. Singh, A., Deedwania, P., Vinay, K., et al. (2020). Is India's health care infrastructure sufficient for handling COVID 19 pandemic. *Int Arch Public Heal Community Med, 4*, 41.
19. Ranganathan, S. (2020). Towards a Holistic Digital Health Ecosystem in India. Obs Res Found April.
20. Rani, R., Kumar, R., Mishra, R., & Sharma, S. K. (2021). Digital health: A panacea in COVID-19 crisis. *Journal of Family Medicine and Primary Care, 10*, 62.
21. Sharma, G. (2021a). What is Right to Education Act (RTE Act)? *India Times*.
22. Kumar, R. P. (2020). New education policy: Five big changes in school, higher education explained. *Mint*.
23. Aissaoui, N. (2021). *The digital divide: a literature review and some directions for future research in light of COVID-19*. Glob Knowledge, Mem Commun.
24. Singh, J. B., Sharma, S. K., & Gupta, P. (2021). Physical learning environment challenges in the digital divide: How to design effective instruction during COVID-19? *Communications of the Association for Information Systems, 48*, 18.
25. Nandini. (2020). PM eVidya programme for digital education in India: Everything you need to know. *Hindustan Times*.
26. Anser, M. K., Ahmad, M., Khan, M. A., et al. (2021). The role of information and communication technologies in mitigating carbon emissions: Evidence from panel quantile regression. *Environmental Science and Pollution Research, 28*, 21065–21084.
27. Daduna, J. R. (2020). Evolution of public transport in rural areas-new technologies and digitization. In: *International Conference on Human-Computer Interaction* (pp. 82–99). Springer.
28. Poliński, J., & Ochociński, K. (2020). Digitization in rail transport. Probl Kolejnictwa.
29. Vermesan, O., Blystad, L.-C., Hank, P., et al. (2013). Smart, connected and mobile: Architecting future electric mobility ecosystems. In: *2013 Design, Automation & Test in Europe Conference & Exhibition (DATE)* (pp. 1740–1744). IEEE.
30. Taumar, D., & Shalini, P. (2020). Only 10–15% penetration of electric cars is expected by 2030 in India. Auto.
31. Kulshreshtha, R., Kumar, A., Tripathi, A., & Likhi, D. K. (2017). Critical success factors in implementation of urban metro system on PPP: A case study of Hyderabad metro. *Global Journal of Flexible Systems Management, 18*, 303–320.
32. Sharma, Y., & Nasreen, R. (2015). Public private partnership in Delhi Tourism—A case study of Delhi Tourism and Transport Development Corporation (DTTDC). *Emerald Emerging Markets Case Studies*.
33. IBEF.org. (2021b). *Financial Services in India*.
34. Kesharwani, S., Sarkar, M. P., & Oberoi, S. (2019). Growing threat of cyber crime in Indian banking sector. *CYBERNOMICS, 1*, 19–22.
35. Chakravarti, A. (2021). RBI plans its own cryptocurrency, proposed crypto law may ban Bitcoins and Dogecoins in India. *India Today*.
36. Nandy, M. (2021). Competition in e-grocery heats up with Tata's entry. *Mint*.
37. IBEF.org. (2021c). *E-commerce Industry in India*.
38. Susan, P. V. (2020). New national retail policy: Licence rationalisation among five focus areas. *Bus. Stand*.

39. Rakheja, H. (2020). Google Launches $10 Bn Digitization Fund to Back India's Digital Future. Inc42.
40. Tandon, S. (2021). Reliance collaborates with micro-retailers to modernise consumption system. *Mint*.
41. Liancre, J. (2020). *Five key trends in digital infrastructure.* Infrastruct. Invest.
42. Dustzadeh, J. (2021). *5 Technology Trends to Impact Digital Infrastructure in 2021.* Equinix.

Conclusion

Infrastructure sector has been a key driver for the growth of the Indian economy. The sector is highly responsible for propelling India's overall development and enjoys intense focus from government for initiating policies that would ensure time-bound creation of world class infrastructure in the country. Infrastructure sector includes power, bridges, dams, roads, and urban infrastructure development. The phase in which India's population is growing needs to have improved transport infrastructure in all the modes such as roads, railways, aviation, shipping and inland waterways. Along with the improved transport infrastructure, India needs modern urban facilities, sustainable power generation and many more. Economic Survey 2017–18 recommends that India requires an investment of over USD 4.5 trillion by 2040 to upgrade its infrastructure. To raise such a huge amount, India needs foreign investment. It should bring regulatory reforms to attract more foreign investment into the sector, including through new investment vehicles and innovative financial instruments. In Union Budget 2021, the government has given a massive push to the infrastructure sector by allocating Rs. 233,083 crore (US$32.02 billion) to enhance the transport infrastructure. The government expanded the 'National Infrastructure Pipeline (NIP)' to 8158 projects from 7400 projects in the Union Budget 2020–21. 217 projects worth Rs. 1.10 lakh crore (US$15.09 billion) were completed as of 2020. Through the NIP, the government invested US$1.4 trillion in infrastructure development as of July 2021.

Acknowledging the importance of the infrastructural development in the economic growth on a country, the book has been conceptualised. This book is an attempt to contribute to this sector by addressing the opportunities and the challenges faced by the industry. This book analysed diverse and complex management issues related to the Indian infrastructural development. It exposed numerous challenges faced by the policymakers and the practitioners in the infrastructural management. This book is first of its nature to analyse sector-wise issues.

Upgrading the Transport Infrastructure

The first section of the book analysed prospects and the problems in the Indian transport infrastructure. All the four modes of transport have been analysed here in detail. Our analysis finds that, despite making so many efforts, India's transport infrastructure is in an abysmal condition. India is far behind compared to other emerging nations like China and Russia. India's road quality is generally low, despite India's roads carrying 90% of passenger traffic and 65% of freight. India's road density is high, but the length of surfaced roads is low at 61% (compared to Russia at 70% or China at 67%). Most of the Indian highways are narrow, congested and poorly surfaced. Though India has launched mega project like the Bharatmala project to boost its road transport network stretching from India's western to eastern land borders, the speed of the work is still not satisfactory.

Soon India is expected to become the world's third largest aviation market after China and the United States and also touch double digit growth rate. Passenger traffic is estimated to be increased by four times and cargo traffic by six times in the next twenty years. Even with this much of potential India's existing airport infrastructure is not efficiently used. Most of the airports in India are underutilised. There are 449 airports or airstrips in India, and commercial airlines are operating at just 61, with the remaining unused or only occasionally used.[1] Many Indian airports have widely recognised deficiencies in areas such as ground handling, night landing systems and cargo handling.

India has unrealised potential in shipping, with 7500 km of coastline and 14,500 km of navigable or potentially navigable waterways. More than one billion tonnes of cargo was handled across over 200 ports in India in 2015 with maritime logistics accounting for 90% of international trade by volume and 72% by value. Despite the cost-efficiency of coastal and inland water transportation, India's ports tend to be small, lack draft for larger vessels and are inefficient.

Indian Railways (IR) is the backbone of India's public transport network. One of the largest employers, it is the greenest, cheapest, fastest, and safest means of surface transport and caters to the needs of both the commuters as well as movement of goods for long distances in India. It is one of the largest networks in the world with 7216 stations; 92,000 km of track and 1.3 million employees. The Ministry of Railways plans to improve and expand the rail network, renew the train fleet and improve passenger safety. To achieve a 5 trillion dollar economy of India by 2024, the Task Force on National Infrastructure Pipeline (NIP) formed by the Government of India (GOI) has projected total investment of Rs. 111 lakh crore in crucial infrastructure sectors over five years FY 2020 to 2025. In railways, projects worth Rs. 13.68 lakh crore (12% of the projected investments) have been identified under this investment package. It plans to invest up to $170 billion over the next five years, with the largest proportion aimed at network expansion and decongestion and safety. Investments are

---

[1] An India Economic Strategy to 2035. Department of Trade and Foreign Affairs. Australia. See https://www.dfat.gov.au/geo/india/ies/chapter-9.html.

Conclusion

also planned for station redevelopment and the dedicated freight corridor between Delhi and Mumbai. The Government of India is seeking greater private investment.

Transition in Energy Sector

Energy is one of the most critical components of infrastructure and therefore very crucial to economic growth. Indeed, it is blood of the economy and crucial input to nearly all the industries. Among all sources of energy, electricity is most visible and is often identified as an indicator of progress in modern civilisation. In India, the demand for electricity is much higher than its supply. India is trying to fulfil its growing energy demands from the alternative sources. It is spending billions in solar power generation. It has taken several steps to reckon with on renewable energy infrastructure in the last decade. India plans to generate renewable energy of 175 GW by 2022 and 450 GW by 2030. Despite several efforts Indian energy sector is struggling with shortage of raw material and coal supplies and inefficient distribution system. Land acquisition became a bottleneck for solar project developers. Energy security still remains crucial as India relied on imports for about 40% of fuel needs in terms of primary energy.

The transportation sector, driven almost entirely on fossil fuels in the form of petroleum products, has a large environmental footprint and linked negative externalities. Not only does the sector impact local air quality and global challenges of climate, but it also has implications on human health and biodiversity. As per IQ Air in 2020 among the world's most polluted countries, India was ranked third in terms of PM 2.5. As per World Air Quality Report (2020), India continues to dominate annual PM 2.5 rankings by city—22 of the top 30 most polluted cities globally are located in India. Traditional modes of transport hugely contribute to this. Realising this crucial factor, India is investing heavily on the modernisation of the transport system and adopting electric vehicles. Transition to electric vehicles can provide benefits like increase in energy security, air pollution and greenhouse gas emissions reduction and industrial development. To adopt electric vehicles and popularise its wider use, India devised National Electric Mobility Mission Plan (NEMMP) 2020. National Electric Mobility Mission Plan 2020 document provides vision and road map for electric vehicle adoption and manufacturing. It was designed to increase national fuel security, to provide affordable and environmentally friendly transportation and to enable automobile industry to achieve leadership in manufacturing.

Revamping Cities

Cities contribute the majority of global economic activity, energy consumption and greenhouse gas emissions. It demands upgradation of urban infrastructures in India. Governments and policymakers are facing challenges such as increasing urban population in rural areas and huge infrastructure gaps. To provide modern and better facilities and to raise the quality of life in Indian cities, the concept of smart city has been devised. The government allocated INR 6450 crore in Budget 2020 to develop five smart cities in the fiscal year FY2021. To accelerate urban infrastructure, the Indian Government set a vision of 100 smart cities starting in 2015. While COVID-19 might have delayed this initiative's progress, the urbanisation endeavour would create jobs and boost economic growth across identified states. These projects require huge upfront investment that could be catered too by accessing builder finance.

Developing Digitally

India has been in the forefront in developing digital infrastructure system. Several policies have been devised to develop and popularise digital services ranging from the financial sector to health sector. Public schemes such as Digital India and creating National Optical Fibre Network are some of the commendable steps taken in India. Digital transformation is now changing the landscape for development. Goods and services can now be unbundled and splintered in global value chains, and they can be transported to anywhere in the world. India can boost up the growth of the sector by scaling up investments in digital infrastructure. It can enhance collaboration with local and global entrepreneurs. It can establish itself as a market leader by expanding its role in the growing global market for digital information-technology services, such as big data and analytics, digital legacy modernisation, climate change agenda and the Internet of things.

Based on the analyses in this book, we may conclude that even though Indian is striving to develop its infrastructure sector, a lot of challenges still remain with varied degrees in different sectors. India needs to address those challenges at the earliest to fetch higher growth.

Rahul Nath Choudhury
Pravin Jadhav